# Some Problems in Information Science

by

## Manfred Kochen

The Scarecrow Press, Inc.

New York and London    1965

[ii]

In Memory of
Max Kochen

## Preface

This book emerged from a two-year study under an Air Force[1] research grant to advance "information science." Intellectual motivation for this study stemmed from puzzlement about the amazing and mysterious ability of people--and perhaps certain automata as well--to understand, remember, recall and process (extract significant items from) vast collections of information at great speed, and in sophisticated ways. Practical motivation came from the increasingly large, complex, and diverse quantities of information people are called on to process, and for which even these remarkable human abilities may soon be inadequate. The root source of this puzzlement seems to be that there is no limitation on the content of information: that these abilities can be as readily applied to one kind of information as to any other.

The primary aim of this book, then, is to help shape first principles of "information science." These deal with precise descriptions of basic functions in a total information system, with conditions for stability and increasing efficiency with growth. Knowledge of such systems can be useful for objectively evaluating proposed schemes of automating various aspects of an information system, for specifying needed developments to attain improvements, for establishing the possibility of a general-purpose, machine-based system. Thus, while primarily theoretical, the work reported here is also concerned with ways of scientific experimentation and the applicability of principles to the specification of real information systems.

The general research strategy underlying this study has been to strike a balance of effort between actual system construction and

3

closely related theoretical and experimental study. "Armchair modeling" in this field might soon turn sterile and irrelevant unless intimately coupled to significant experiments. To do the latter with any kind of realism, however, required a very powerful, non-existent experimental facility of considerable flexibility. This would involve a real user community to be used as experimental subjects, yet offering the possibility of some experimental control, and a critical size corpus of information containing answers to queries posed by the users. The attempt to obtain such an experimental tool resulted in the specification of the AMNIP system and the construction of a modest system which resembles AMNIPS.

A large number of investigators from different scientific disciplines contributed directly to this study. We wanted to present a sample of thinking in this field which represented the entire area rather than a special subarea, realizing that such broad coverage must be sparse and incomplete. Common to all our varied interests and backgrounds was a desire to bring rigorous and creative thinking from our training and experience in both the exact and the behavioral sciences to bear fruitfully on the following general problem: How can relevant information in the increasingly complex environment of an organism be internally represented, stored, processed so that increasingly effective actions (and increasingly accurate predictions) will result? And in the sophisticated use of modern computers and concepts in the computer art we saw a possible tool not available to our predecessors, supporting our hope that we might make progress in some areas in which they did not.

In order to attack so vast a problem, , with relatively little firm foundation upon which to build  a few points of attack had to be selected. That this choice was judicious in spite of many false starts can be judged from the level of understanding and the specific results which were attained. The technical results are reported as separate papers written by the authors directly responsible for them. All these papers should be regarded as reports of work in progress, though a few of them may prove to be of value as they stand, or stimulate the future production of results of even greater value. The separate papers were all assigned under the general plan of the sponsored research program. There was no attempt to screen papers according to a single criterion. While the more mathematical papers might conform to standards of rigor, the papers which formulate problems and discuss issues might be of help to those seeking order, structure and clarification of what "information retrieval" is about.

For these reasons, this book might prove useful to graduate students with some technical background and interested in this interdisciplinary field; to library scientists wishing to sample the kind of thinking that might have impact on their profession, to engineers and scientists wishing to contribute to the further development of "information science." If it succeeds in provoking thoughtful examinations in directions that had previously not been recognized as significant and in stimulating further contributions in those directions, its aim will have been met.

December 30, 1964                                      Manfred Kochen

### Note

1.  The support of the Air Force Cambridge Research Laboratories, Electronic Systems Directorate, under contract AF 19(628)-2752, with the IBM Research Center is hereby gratefully acknowledged.

Table of Contents

10

# I. Introduction

<u>Background</u>

The study reported in this book began in 1961. It followed a comprehensive, critical survey of the literature relevant to "artificial intelligence." It was also based on a computer program for testing reception strategies in the attainment of concepts and revision of hypotheses. The survey had left us with the conviction that goals and standards in this field still required formulation, and that the greatest lack was still of new[1] concepts and ways to evaluate them relative to real problems. Computer stimulation of hypothesis-selection proved itself a useful tool for scientific experimentation as well as for theoretical study of new systems concepts on which to build a study program in "information science."

The current approaches to machine "intelligence," including our own, dealt, more or less efficiently, either with problems of a special, sharply delineated part of the real world or an abstract situation, while the key task for cognitive machines was, rather, to learn to deal with any aspect of the real world. Study of the mechanisms of language, as related to the formation of internal models of the relevant real world, seemed the best way to proceed. But "computational linguistics" with its stress on syntax offered a no more fruitful approach for our purpose than did game-playing, theorem-proving, and nonlinguistic symbol-manipulating tasks based on heuristics. For this reason, a new approach was developed that emphasized: (1) continual adaptation to the relevant real world, (2) division of labor between human and automatic elements of an information system with effective (linguistic) commmunication between them, (3) methods for automatically representing the relevant real

11

world as internal maps to guide decision making.

Thus, it appeared to us, as to many others, that "information retrieval" was a good vehicle to study "artificial intelligence," or to use a better term, cognitive processes. Though the first area was hardly better defined than the second, there were real problems[2]--in automation of libraries, special information centers, in the publishing industry--and real opportunities to attack them.

On the intellectual side, two documents had a major influence on our thinking: V. Bush's Memex proposal and the Weinberg report (even before it appeared, because our thinking was, independently, along identical lines). On the practical side, we tried to steer a course between the two opposite poles that dominate information retrieval research today: at one extreme, a tendency to seize on some particular idea for a system and mount a crash program "to get IR moving," to demonstrate a working and useful system; at the other extreme, to do careful systems analyses instead of, or prior to, "trying out ideas" by expensive system construction efforts. We soon realized that there is no adequate intellectual foundation, no relevant underlying scientific discipline with either approach when it comes to more sophisticated information systems problems.

Orientation

While we strongly favor quantitative analysis of proposed means for improving information systems in relation to alternatives over partisan support of a particular system, the lack of established principles made it temporarily expedient to adopt a position, based on the best available judgment rather than facts. This position was a compromise between that taken by enthusiasts for fully automatic indexing, abstracting, content-analysis of text in "natural" language and servicing queries posed in unconstrained "natural" language at one extreme, and the position taken by enthusiasts for rigidly formatted special inputs and highly constrained programming languages for posing queries at the other extreme. The AMNIP (adaptive man-machine nonarithmetic information processing) system is the result of such compromise.

This system was conceived and specified in 1961. Its implementation proceeded as a one-man effort. It was implemented on a
novel hardware system, which is still in an experimental stage.
This hardware system combines a photoscopic disc storage unit with
a 7094 computer and a special console for on-demand access to a
content-addressable, large, high-speed random access memory.
The programs were written to permit time-sharing.

The main claim to be made for AMNIPS concept is its potential
usefulness as a "general-purpose" experimental tool. A brief description of the AMNIPS idea is contained in I A. This forms the basis for
many of the other papers in this book, notably II D, II E, III B, III C,
IV A, IV C.

The prior work on concept-learning, of which AMNIPS is an out-
growth, is directly related to the papers II B and II C. The underlying
sociological conditions for success of an AMNIP-type system are discussed in papers I B, IV C, IV D and IV E. Some more detailed technical considerations in the design of an AMNIP-type system--coding of
names for convenient access, storage of interlinked names, systems
analysis--are discussed in papers I D, III D, III E, IV F.

Only the papers II A, III A and IV B are not directly connected
with AMNIPS. The latter is a by-product, another application of
algorithms for grouping similar names in AMNIPS. Papers II A
and III A deal more speculatively with questions underlying the use
of information systems generally.

The goal to which this study was directed was solution of the
following general problem. How can relevant information in the increasingly complex environment of an organism (modeled as an automaton) be internally represented, stored/recalled, and processed
efficiently so that increasingly effective actions can result? Toward
this objective we attacked simultaneously along three fronts. The
first was to investigate the overall aspects of information systems,
with special emphasis on the role of automation in amplifying certain human cognitive abilities--e.g. understanding, knowledge (Chapters I and II). The second was to examine the type of memory

organization best suited for this, with special emphasis on "associative" memories, list-structures and key-address transformations (see Ch. III).  The third was concerned with how such a system grows in its ability to help solve problems and process data relative to its rate of growth in recorded facts (see Ch. IV).

The twenty papers in this book show how computer simulation based on list-processing, theoretical discussions drawing on sociology and the psychology of cognition, engineering studies of semi-automated library systems, and mathematical analysis applying graph and automata theory, all were used to attack various aspects of the central problem.  In particular, they report some progress toward classification of types of discourse; toward modeling of an information system in terms of memory, processor and "comprehender" subsystems governed by self-regulatory principles; toward operational performance analyses of certain special information systems.  Computer programs for concept and language learning and of a general-purpose experimental information system storing information as sentences composed of names and predicates (AMNIPS) were specified and constructed.  A critical survey of associative memories, theorems about new and improved key-to-address transformations, about a hardware-software compromise to multiple descriptor record retrieval, as well as algorithms, and programs based on them, for grouping items with similar attributes will also be found.  In all, these reflections contribute a beginning toward information science.

Information Systems

"Information science" does not now exist as a scientific discipline in its own right. [3]  We believe that it may emerge as such, perhaps as a synthesis of concepts and methods originating in automata theory, logic, mathematical linguistics, graph-theory, list-processing techniques, and the theory of cognitive processes.  It deals with the principles and facts underlying the analysis and design of information systems.

We interpret the information system of an organism to have

the functions of planning[4] the behavior of an organism, of alerting the organism to changes in its environment that signal the need for action, of controlling action toward the implementation of a plan.  To function thus, an information system should continually form and reformulate an internal representation of the organism's relevant external world to make it capable of increasingly effective actions (and predictions).

We view an abstract information system to have three major subfunctions: an information storage and recall facility; a problem-solving facility; a facility for comprehension,[5] representation, integration of cumulated information.  These correspond, for example, to the subsystems of the information system of an institution, like a university: libraries are usually responsible for storage and recall; their user community is responsible for solving problems and processing the information retrieved from libraries; the truths and techniques known to this community and the jargon that goes with them constitute the knowledge created and used to add up to understanding.

## The Library or Storage/Recall Subsystem

The first subsystem is a collection of documents and means for filing, searching and retrieving.  It is a library.  What is crucial about a document is that it authenticates, for reliable use by posterity, an item of information in a respresentation suitable for the type of communication intended.  It is an author--the one who has intellectual responsibility for a document--who plays a key role in authenticating a document.  Referees, editors, publishers, reviewers, all play a part, of course.

One of the key relations among documents is that of citation. Through such links, the network of documents that make up a library is organized into a nonhomogeneous structure composed of local document groupings or clusters of "similar" documents.  Citation links are important because they trace the explicit origins from which a document was synthesized, and they indicate for which later documents a given document serves as ingredient.

In Chapter III, this subsystem is discussed in terms of a few
ideas and problems dealing with hierarchies of documents organized
into an encyclopedia, with citation nets, with ways to organize and
address a file of documents, with associative memories.

## The Data-Processing Subsystem

The second subsystem is a collection of (human, possibly
aided by machines) agents in roles concerning generation of information (e.g. authors), primary and secondary publication, problem-
solving, data-processing, facilitating search by finding unsuspected
connections, patterns among data.    Such an agent functions by virtue of his interactions with other agents, by selective exposure to
literature, by reasoning and reflection.    One of the principal tasks
of certain agents is to plan; another is to stay alert to current information in order to assess trends, deviation from trends, deviations from a planned course of action requiring corrective action,
in order to reformulate goals when necessary. [6]    To do this, timely
and convenient availability of relevant information--both that known
to exist and called for, as well as that not known to exist but that
would have been called for had it been known to exist--is vital.
The process of formulating goals, analyzing a goal into subgoals,
then deciding whether information needed to attain a subgoal is to
be sought by recall (i.e., lookup, search) or analysis of the subgoal into further subgoals is fundamentally similar to the process
of problem solving. [7]    What is significant for information science,
however, is that the agents engaged in these activities function by
virtue of interactions among themselves.    These interactions may be
links of intellectual influence among the agents, viewed as modes of
a large, nonhomogeneous graph.    This graph, like that of the interlinked documents in a library, can thus be organized into "clusters"
of people, corresponding to cliques or "invisible colleges."

In chapter IV, a few of these ideas are examined further,
notably that of a dynamic net of communication channels among professional colleagues, and techniques for grouping similar items so
as to facilitate search and problem-solving.    Classifying information

items so as to match the way human problem-solvers might classify
them may provide short-cuts to exhaustive searching in many problem-solving and data-processing tasks, much the way heuristics do.

## The Knowledge Subsystem[8]

The third subsystem is a collection of concepts, percepts,
ideas, facts, findings, judgments, explanations. These are the raw
materials of thought, the units of cognition, with which the information process (from generation to use) is concerned. As a document
has both an abstract aspect--a book is really an abstraction, for
when we speak of "Ivanhoe" we don't mean any particular copy but
that which is common to all copies of the work--and a concrete as-
pect--e. g. a particular copy located on a given shelf at a given
time--and, as an agent has both an abstract and a concrete aspect,
so do the units of knowledge in this third subsystem. The concrete
aspect of a fact would be the actual event or state that is; of an
idea, the neural activity associated with it. The abstract aspect of
a fact is the linguistic or other form in which it is represented.
This is not the same as a document, because these units of knowledge, though they may be represented in a form superficially re-
sembling that of a document, are not necessarily documented, au-
thenticated. Only when the process of transforming such knowledge
into documents is complete are they in form suitable for lasting
preservation.

These units of knowledge could be viewed as nodes in a graph,
concretely represented by names, subject headings, etc. (e. g.
Boyle's law, the Whorfian hypothesis, the concept of chirality, the
idea of a "World Brain"). These are interlinked through such links
as implication and relevance. Again, inhomogeneities in the graph
may reveal a macrostructure of "clusters" of significantly interrelated items.

This way of looking at an information system and its relation
to ideas used in AMNIPS, is described in I C. It casts "information science" clearly into the behavioral sciences. It stresses problems of stable growth as central ones. It sets the stage for

construction of more quantitative, yet profound, models and subsequent analyses of information systems. Yet, in its present form, it is quite far from the needed logical precision and quantification. This will be done in a separate volume.

Paper I D illustrates the kind of quantitative information systems analysis that can now be done, at the expense of the intellectual depth in the above, less precise, formulation. The particular model which is presented deals with a special type of information system: a computer-based, on-demand, coordinate-indexed system which furnishes bibliographies in response to requests by topic. It does not include the practically more frequent document-fetching service, nor retrieval requiring inference (type C), though these are now being incorporated. It places undue stress on hit-rate and acceptance-rate as measures of performance but more profound performance variables are now under investigation. Net profit is used as a figure of merit for the entire system because price is one of the few scales along which the various factors in a systems analysis are comparable. Conditions on the system parameters (e. g. average random access time, the minimum number of users) for profitability are derived. It does specify the kind of information needed from empirical and experimental work. It also illustrates the kind of systems arithmetic which very badly needs doing--and systematizing.

In relation to the overall view of an abstract information system presented in this chapter, the indexed document collection corresponds to the "library subsystem" with the storage/recall function. The collection (or authority list) of index terms, which may be the bulk of the vocabulary and phrases in the texts of the collection, and the various interrelations among them, constitutes the "knowledge subsystem" with the representation function. The querists in this model sketched in I D, parametrized by the number of terms/query and the number of near-synonyms/term for this simplest model, form the "processing subsystem. "

### Notes

1.   In this frontier field, very many seemingly good and new concepts present themselves.  A major difficulty lies in applying judgment to extract the genuinely good ideas prior to expensive operational tests.

2.   These information retrieval problems may, for purposes of preliminary classification, be grouped into three categories, according to whether they deal with the analysis of systems for answering queries of the form:

A.   Do you have document X, which is known to exist, and what is its present availability status?

B.   What documents, if any, exist on a topic on which information is needed?

C.   What is the answer to the question like "What is the maximum temperature on the surface of the moon," the answer to which is known not to exist explicitly.

3.   We question the view that it could never be a discipline but only an application area, applying results and techniques from established disciplines.   There is not enough revelant knowledge in these disciplines to apply.   "Information science," as an emerging discipline, may more likely revitalize, direct, stimulate the older disciplines than vice versa.

4.   This thesis is developed more fully in II A.   This paper studies planning and alerting process in more detail, with a view toward a descriptive model.   If, with the help of such a model, we could specify conditions for instability and malfunction, we could recommend the introduction of automation in certain functions as possible remedies.   The AMNIPS idea, like its predecessor, the MEMEX idea, may lead to such a remedy.

5.   We have not yet found the single most appropriate term for the precise nature of this function, with which we have yet to come to grips.

6.   This point is developed further in I B, where the self-regulatory control mechanisms which govern the functions of an

information system are discussed.    An understanding of these mechanisms is essential for understanding how planning and alerting
could take place.

7.    This formulation, tied to the Newell-Simon GPS, permits
a fresh approach to the problem of how to (normatively) prescribe
when to search the literature or to resort to original analysis in
solving problems.

8.    Developed as chapter II.

I A.  An Adaptive System for Directly Recording and Retrieving
Information in Simple, Formal, English-Like Sentences[1]

M. Kochen

Abstract: A system, consisting of a general-purpose IPL-V program,
to run on a novel computer-large store-console complex, for recording and retrieving information presented in simple, formal,
English-like sentences, through direct man-machine coupling, has
been constructed.  It applies to situations where generation of information is subject to some control, as in registries of data about
projects or people.  Two key features of this system are:
1)  it should improve, with use, in its ability to respond to queries
as well as to assimilate new entires;
2)  it should invite use by convenient man-machine coupling.

Information systems can be grouped in two major classes, depending on whether generation of information is subject to control or
not.  The first kind of information system would typically occur in
a registry of projects or people, where an unspecified variety of
facts is to be recorded in order to answer queries and make com-
pilations.  Such information systems, to regulate the flow of certain
messages, would also be found in some early warning systems.
The formal language in which such data are to be recorded directly
by the originators should be: a)  of sufficient expressive power to
allow eventual representation of the kinds of data likely to be recorded; b)  sufficiently precise to allow machine inferences and
searching; c)  sufficiently similar to natural language to make it as
convenient as possible for users to learn it.  The use of such a

language will gain acceptance much as the use of a typewriter or of a telephone dial gained acceptance over prior manual methods of performing the same task, or over such proposed alternatives as automated recognition of handwriting or voice.   The user has more to gain by way of automation in return for the minor readjustment (where this is feasible) of his mode of expression to suit the machine than we would by having the machine designed to accept his greatly varied "natural" language expressions.

For systems where such control of information generation is feasible, a simple formal language of quite general applicability has been designed.   A programmed system based on this language, to be used on a machine complex consisting of million-bit millisecond random-access stores, coupled with 7090-type computers, coupled with consoles, for "on-demand conversational interaction," has been constructed.   The data base for a trial demonstration as accumulated so far consists of about 100,000 facts contained in the booklet Current Res. and Dev. in Documentation, No. 10, by NSF.

Sentences of the formal language are built of names and logical predicates, which can be chosen to suit the particular application.   To record biographic data for example, there are predicates like, "_____ was born in the town of _____," "_____ was affiliated with _____ in the position of _____ from year _____ to year _____," "_____ was educated in the field of _____ at _____ from _____ to _____." By choosing the predicate "_____ cites _____," a citation net can be obtained (see IV B); by using "_____ is a near-synonym of _____," a type of thesaurus or synonym dictionary (III D, III E) is obtained.   Names of people, places, dates, institutions, are inserted in the blanks.   Programs for accepting such sentences via console or card reader, updating a cumulative store of such sentences, and perhaps storing certain references like "If X is author of Y, and Y pertains to topic Z, then X has worked on topic Z," have been specified.   Programs for updating and interrogating this fact file, for tracking certain trails among the net of names thus linked through various predicates have been written in IPL-V. Thus, whenever a sentence like "Doc. a cites doc. b" arrives, the sentence, "Doc. b is cited by doc. a" is automatically formed and stored; if a querist wants information about doc. a, the machine would display to him, at the console, the variety of predicates he could use to revise his query and make it more specific, e. g. what documents cite document a.

Such a system is envisaged to be used by project leaders, sponsors, administrators, survey-writers, authors, reviewers, who would directly enter data or queries through consoles or reactive typewriters. This would be an automated version of what could, at first, be implemented by having the registrants fill out a check-list questionnaire, which provides the system with some preliminary clues about the nature of the respondent's entry or query; tailored to the response, a more specialized check-list questionnaire--in the form of the sentences described earlier, and eventually displayed on consoles or printed by remote typewriters rather than mailed--continues until the system's and the man's levels of informedness are matched. A key question is to speed the convergence of this process so that it becomes practical to use.

While such a system will not suffice where input information is historical--e. g. archives, old books and papers, statutes--or in intelligence problems where control over relevant currently generated information is clearly out of the question--e. g. newspaper reports, transcripts of a patient's verbal behavior in a psychiatric interview--there are applications where information generation can be standardized. Important examples other than biographic and project registries are bibliographic control (e. g. man-machine cataloging), recording of certain new scientific and technical data, documentation of certain current events based on first-hand observation. This system goes beyond conventional methods of processing formatted data in that it is more amenable to experimentation with improved automation, inferences, searching, representation of stored data, and updating.

The two key features of this system are that it should improve, with use, in its ability to respond to queries as well as to assimilate new entries, and that it should invite use by convenient man-machine coupling.  It is to do the former by ensuring that the mean time to locate a needed name, or fact about a needed name, by having trails in the net of names interconnected through predicates, does not grow faster than the total number of sentences in storage. This is to be done by procedures for grouping (classifying names with "similar meaning") as the store grows.

The AMNIP system is described in greater detail in the article, "Adaptive Mechanisms in Digital Concept Processing, " by M. Kochen, Proc. Symposium on Discrete Adaptive Processes, June 1962, pp. 50-59.


## Notes

1.   This work was supported in part by Air Force Contract AF 19(628)-2752.   The paper is published in Automation and Scientific Communication, Proceedings of the American Documentation Institute, 28th Annual Meeting 1963, Vol. 1, pp. 43-44 and is reprinted here with the permission of ADI.


## References

Bush, V. "As We May Think" Atlantic Monthly, July 1945, pp. 101-108.

Hunt, E. A.  Concept-Learning, J. Wiley and Sons, New York, 1963.

Stevens, M. E.  "A Machine Model of Recall" Proc. Int. Conf. on Information Processing, UNESCO House, Paris 1959, pp. 309-315.

I B.   Total Information Systems in Planning and Alerting

M.  Kochen

Lecture presented at the Moore School of Electrical Engineering,
University of Pennsylvania, in Monthly Seminar on Automatic
Computers and their capabilities, October 2, 1963[1]

1.   Current Concern with Information Retrieval Rooted in Malfunc-
     tion Danger of Planning and Alerting

     Last June an inventory of science information activities in the
Federal government was published.   It emphasized the urgency with
which science information problems were being recognized and attacked.   It followed in the footsteps of the 1962 Weinberg report[2]
in which the central problem facing the national community in man-
aging scientific research and applications was defined in terms of
the danger of science fragmenting into a mass of repetitious findings, or worse, into conflicting specialties that are not recognized
as being mutually inconsistent.

     Thus, there are now over 470 special information centers in
the U. S. alone.   Over 350 research projects are investigating information retrieval or related subjects.   Twenty federal agencies
with significant scientific and technical information activities have
each established a focal point of responsibility, and all these are
coordinated into the Committee on Scientific Information of the
President's Council for Science and Technology.   The National Bureau of Standards is developing a system to provide scientists and
engineers with critically evaluated numerical data in the physical
and engineering sciences necessary for the research and development

process.   The Science Information Exchange, now under NSF spon-
sorship, is extending its coverage to include projects in progress
in the physical sciences as well.   The Library of Congress, with
the NSF, has established a system to direct any scientist to the
source of information most appropriate to any scientific or technical
question.   Nonetheless, the government is still urgently concerned
with improving the rate at which technological advances achieved
through its ten billion dollar per year R & D expenditures are recognized and adopted by business.

The danger of a professional activity fragmenting into repetitious or incompatible specialties is by no means unique, or even
most critical, in science.   The danger is present in the medical,
legal, political, and economic professions as well.   But what, precisely, is the danger?   To understand this is to understand the
functions, and possible malfunctions, in the information and control
processes responsible for keeping a professional activity integrated.

Three essential functions of management in any organization
are: planning, organizing and controlling.   The great current concern with information is rooted in the difficulties of effectively man-
aging modern giant organizations which are concerned with such professional and other activities.   These difficulties, in turn, stem
from increased information output which is itself a result of population expansion and organizational growth.   Each year, over 25
billion pieces[3] of paper are created by government alone.   The
world's scientific literature output has been estimated at 1. 2 million
papers a year.   About 450, 000 papers a year appear in standard
American technical journals, and these are now generally abstracted
and indexed; also, under bibliographic control are the approximately
100, 000 books a year catalogued by the Library of Congress.   But
about 100, 000 government reports[4] are often not indexed, and these
are playing an increasingly important role in technological planning.

By planning, we shall mean the continual formulation of goals
and the guidance of action toward their attainment.   To develop the
Bell Telephone system, for example, required comprehensive

planning.  When an individual establishes a research project to satisfy his curiosity about a challenging puzzle, essentially the same
planning processes are at work.

A concrete example of a plan, for instance a corporate long-
range plan, is a 100-page document containing financial schedules,
growth charts, and other precise statements.  On the first page is
a statement of objectives, which may be:

"To diversify our product line so as to minimize dependence
on X products for existence and growth.  By 1968, at least 20% of
earnings should be from sources other than X products. "

A plan succeeds if it leads to actions which effectively reduce
the needs that prompted it.  The success of a plan depends critically on rapid, convenient availability of complete, accurate and relevant and only relevant information.  To inform the planner what
available information he ought to have is more important than supplying what he asks for.  He must have the information he needs,
not just what he specifies, in order to continually reformulate his
goals toward greater precision and toward mapping an effective
strategy.  And he often does not know what he needs until he almost
has it.

A key function in implementing a plan is to continually monitor the environment for signs of unusual change so that timely responses can be made.  An alert to such change is successful if
events likely to prove significant can be signalled as such, and in
time.  If too much current input information must be scanned in a
short time relative to the time needed for digesting, assimilating
and evaluating such information, then alerting will malfunction.

If there is malfunction in either alerting or planning, the underlying human needs will continue unreduced and may find explosive
outlets.

Now what are the danger signs that planning and alerting
could malfunction?

Small corporations--those with less than $1 million annual
gross sales--which plan developments counter pending legislation

may suffer losses to the extent of being affected by 1 bill/year.
Time lags in reporting statutes are 1-1/2 to 2 years, and information
about pending legislation is not conveniently accessible to small cor-
porations.    Similarly, pharmacists find it difficult to keep up with
regulatory actions by the Food & Drug Administration, and there is
no adequate way to alert drug manufacturers and doctors to the latest results of current research and tests involving new drugs.

A more humorous example of poor planning due to difficulties
in getting pertinent data, or rather due to lack of knowledge about
what data to get and how to use it, was manufacture of the hula
hoop.   The number made exceeded the number of children in the
U. S., though the life cycle of this product could have been estimated to be about 120 days.

Inventors and technologists are in danger of finding many of
their efforts anticipated somewhere in the enormous literature.   A
Soviet journal in 1961 reported that 80% of the 130,000 new inven-
tions reported to the USSR government were found to have been
published or patented previously.   This does not imply that technologists should spend most of their time digesting the literature.   The
Weinberg panel found that overall productivity is greatest when time
is equally divided between trying to create new data and digesting
other work plus communicating one's own.

These dangers for malfunction in planning and alerting have
always been present.   And, as Ralph Shaw has pointed out, there
are excellent search tools like road maps, encyclopedia, handbooks,
indexes, poison control centers and good current awareness tools
like "Science Newsletter," newspapers, etc. which, in the past,
effectively guarded against such dangers.   What makes them critical
today is that the rate of information generation may soon exceed
the rate at which information is integrated into a genuinely coherent
whole.   This encyclopedic task of assimilating new information is
beyond the primarily custodial responsibility of the library profession.   It can be accomplished only if enough of the most capable
professionals will compact, review, interpret and synthesize the

literature of their profession.

## 2.   Planning and Alerting Based on Models

This task of integrating knowledge is of great intellectual challenge and responsibility.   It should be entrusted to the very best, of whom there are few.   It is, therefore, reasonable to examine what automation might offer to increase the productivity of such people, so that one could do the work of a hundred. [5]   To explore the potential of automation is not only to assess trends of very impressive recent developments in technology; it is even more important to understand the precise nature of the task to be performed by the encyclopedists to help in planning and alerting; to determine which of these tasks need aid; and when this aid should be in the form of automation.

It is vain to look to automation to do the job of the encyclopedists.   I say this because sifting, reviewing, synthesizing all that is known involve not merely processing information with regard to its cognitive aspects.   That is, switching of information and even deductive inference according to rules of logic--without regard to its meaning and significance--is not enough.   The motivational and evaluative aspects of handling information must be taken into account as well, and no mechanical procedure has, even in principle, so far been demonstrated capable of creating new knowledge the way a scientist would by the irrational process of discovery and synthesis of the known into the new.

Of course, such a mechanical superscientist might be nurtured into existence, if the dreams of the more enthusiastic exponents in artificial intelligence are realized.   Indeed, five of us spent five weeks at RAND last summer examining the differences between human comprehension and its machine analog.   The one conclusion to which we all agreed was that, while there seemed to be no theoretical barrier to the existence of a machine which exhibits comprehension, this would require the integration of a wholly distinct set of capacities with those embodied in existing programs.

A more realistic goal for research in artificial intelligence in this context would be to study the process by which understanding is acquired and how it is used in planning and alerting and, based on this, to specify a division of tasks between top encyclopedists and their mechanical assistants so as to maximize total productivity.

Let us take a look at planning.   Planning begins by recognizing a need, even though this may at first be vaguely and dimly perceived.   This is followed by groping for suitable concepts and goals which are formulated and continually reformulated with increasing precision and accuracy.   For help in this process, the planner interacts with colleagues, searches the literature and reflects.   The most important thing that happens during the interaction between the planner and the work of the encyclopedists is a cultural transfer of a relevant model as they have synthesized it from all that is known. All further planning is based on such a model.   In the process of transfer, the planner may, of course, enrich the model.

A model is an idealized representation of the real world which can be used to guide action.   Sometimes these representations are explicit in the form of linguistic utterances, pictures, or abstract symbolic logic; sometimes they are implicit percepts, concepts or ideas inside the planner's mind only.   It may be of interest to ask if anything that is imaginable--that is, every implicit model--can also be explicitly represented by means of symbols.

A large class of models could be usefully thought of in terms of graphs.   The nodes stand for representations of states in the real world.   The edges stand for allowed transitions from one state to another through time.   Certain section graphs of this graph are also represented as nodes in the graph.   This makes it possible to assign to many pairs of nodes a third node in the graph so as to interrelate the nodes into at least a monoidal structure.   Two models can then be related.   One might be a homomorphic image of the other.

The nodes of the graph for an explicit model could, for example, denote board configurations in checkers or Boolean propositions, as in the Newell-Simon General Problem-Solver. There, eight rules of inference specify allowable links from each node to eight others. One node is selected as a goal-state by the planner, and one as his starting state. If the shortest path from the starting state to the goal state is known to pass through n intermediary states, then as many as $8^n$ paths might have to be examined in the worst case of an exhaustive search. For this reason, some of the intermediary nodes are chosen as fixed stopover points and are interpreted as subgoals. Just one stopover point halfway between starting and goal state would reduce the number of false trails, in the worst case, from $8^n$ to $2 \cdot 4^n$. The judicious or fortunate choice of such subgoals is not to be derived from searching but from the examination of a homomorphic model or from an analogous model with which there is greater familiarity.

For the case of explicit models, the processes of planning and problem-solving have much in common. The simplest kind of problem is to check whether a given deduction is consistent with the rules of inference. Visualizing the model of logical states as a graph again, a proof is a path leading from a node representing the first proposition to the node representing the conclusion. Some of the intermediate nodes are terminals of branching paths from nodes starting with the premises to nodes used as steps in the proof. Checking whether rules of inference are correctly applied is simply checking that no forbidden links among nodes have been used.

A second kind of problem is to find a derivation leading from given premises to a given result. In deductive logic, there is, of course, no guarantee that a path can be found according to some rule which uniquely determines each step in the search, and such that the search terminates in a finite number of steps. In terms of the graph representation it can, perhaps, be shown that there is a probabilistic search strategy for which the mean search time is finite under general conditions.

The proof-finding problem has the following important analog in inductive logic: given factual data and an hypothesis, to what

extent does the evidence confirm it?   Again, the graph picture helps
to visualize the situation.   The edges no longer stand for permis-
sible links of inference but confirmation or refutation by evidence.
Large accumulations of confirming evidence would appear on the
graph as clusters of nodes representing states of the world centered
on a node representing the hypothesis.   Decomposing a graph into
several such clusters, each centered on the node for a competing
hypothesis, at least a partial ordering on the clusters can be found.

   The third and most interesting type of problem is that of de-
riving from axioms a theorem which is interesting and, perhaps, related to some linguistically specified topic.   The inductive counterpart of this problem is to conjecture a hypothesis to account for
given empirical data.   For both of these problems we no longer deal
with explicit models, where the nodes of the graph can be represented by symbolic expressions.   These problems are closely related to the key problems of formulating goals which the planner with
a vaguely felt need must face.

   Having formulated a goal and a path to attain it, the planner
begins implementation.   To ensure that a plan is properly implemented, much of the environment must be continually monitored.
To the extent that the plan is based on a powerful and realistic
model, a number of indicators can be specified and monitored.   If,
for example, the model is of a mathematical statistical type, some
of the indicators might be random variables, such as the number of
freight cars per hour passing through Erie.   If the value of any one
such indicator, or certain combinations of indicators, falls in a critical region, there is an alert signal.   The important thing about
such an alerting process is that the probability of no alert in case
of true danger or opportunity, as well as the probability of false
alerts, be low; and that the alert signal arrive in time for appropriate action to be taken.

   The key conceptual problem here is how to calculate the probability of no alert in the case of genuine danger or opportunity.   If
the underlying model of the real world has been sufficiently precise

and accurate, and properly used relative to a precisely specified goal, all possible situations of danger or opportunity would have been foreseen and taken care of in the choice of indicators. But such is never the case, particularly when situations are being modeled that involve people, where evaluative and motivational factors play a role. Hence, no set of indicators can be exclusively relied on for efficient alerting. In addition, the great mass of unstructured current information must be monitored as well. Since no specific events to be looked for can be a priori specified, some expected or normal state must be attributed to the environment and all significant departures must be watched for. These are then correlated into much more comprehensive models than the one on which planning was based. The unusual or anomalous event in isolation is never revealing, but attempts to group several such events may re-result in revealing patterns and cause an alert signal.

While monitoring by following indicators is a form of hypothesis-testing, monitoring of unstructured current information may lead to new hypotheses. It may enrich the model, cause a reformulation of planning goals.

An important theoretical issue is how much effort to expend in monitoring indicators relative to monitoring unstructured intelligence. Some indicators--for example, the machine tool index, which is regularly compiled for economists--have been judged of little predictive value and need not be monitored. Since we cannot set bounds on how much of the unstructured data to monitor--because significant events could and do arise in the most unexpected quarters, particularly in dealing with an adversary--the vast mass of unstructrued data must be sampled. For a situation in which a good model can be constructed--where there is no unpredictable adversary, no motivational or evaluative aspect to be dealt with--in short, something like a physical situation such as tracking a ballistic missile--most of the effort should go into monitoring indicators.

3.   Automation Can Help

Now we might ask which of these functions could benefit from automation, how, and how much.  I will now present two examples of how automation has helped in some of the tasks faced by the encyclopedists and the users of their product.  These automated tasks are peripheral to the central tasks, on which the value of automation has not yet been demonstrated.

First, there have been impressive results in automatic composing.  The Los Angeles Times has displaced 45 positions in its composing room beyond the 10-position displacement required to break even for the use of computers in typesetting alone.  The most impressive accomplishment, from an intellectual[6] point of view, is automatic hyphenation of nonhyphenated words to aid in justifying the right-hand margin of a news column.  The remarkable thing is that this is done primarily by logic rather than by relying exclusively on table lookup.  As a result, dividing the average word takes only 15 milliseconds of machine-time.  Of the 428,000 lines of print that have been hyphenated in the first 7 months of 1963, only 580 words were found to be in error.  Of these, 460 were corrected so that as of last July, the hyphenation program was good for 99.7% of all cases.

The significance of this accomplishment is to demonstrate that some of the routine processing on the text of news stories, which arrive in machine-readable form as teletype tape, can be advantageously automated.  If this can be successfully done, why not cumulate several years' worth of the New York Times Index or the index to the very comprehensive files of articles by Time, Life, and Fortune?  Perhaps the index, which contains cryptic sentences summarizing news stories, could also be prepared by machine.  The resulting cumulative index, together with the library of the news stories themselves, if conveniently accessible, would probably be a very useful and comprehensive encyclopedia of current history.  As such, it could help the encyclopedists in their task of integrating knowledge as well as the planner in his alerting task.

The second example illustrates the advantages gained by close man-machine conversation and suggests how similar techniques might help the encyclopedists and planners in their task of assimilating past literature and in retrospective searching. The example is again in a peripheral area, this time in the area of teaching machines.

An experimental computer-based teaching system, [7] to which several students have real-time access, has been used to teach 40 students from a small college a programmed one-semester course in elementary psychological statistics. After a median of 12 hours of interaction with the computer, through reactive typewriters, and with no time spent on extra homework, or reading, the students attained a median grade of 87% on a final given for the equivalent course at the college. The impressive thing is the reduction of instruction time. A whole one-semester's course in a long day's worth of instruction! The major reasons for this are three: the student gets immediate reinforcement due to the on-demand access to the system; the student is active rather than passive, because the machine responds and demands further responses from him; the course programmer is forced to organize his material with the precision demanded by a machine system. These same reasons for advantageous automation may apply to situations in which the person initiates the question in order to be informed instead of the machine initiating questions in order to inform.

While these accomplishments are impressive, they are in peripheral areas. No such accomplishments could be claimed for automation in the more central tasks of helping encyclopedists and planners to cope with the literature. Consider how effectively a completely unautomated specialized information center served the appropriate scientific community, namely the one developed by H. Selye's at the University of Montreal. This collection consisted of 400,000 reprints, journals, books on stress and endocrinology. It was partially destroyed by fire in 1964. About 200 literature searches were made each month. The mean time to search the file was about one minute. This included the time for processing

the request and to search the files.   The file grows at a rate of
20,000 pieces a year.   Cataloging is reported to take only 5 minutes
of clerical time per item.   All this is done by a staff of six coders
trained in physiology and nine filing clerks at a cost of $10-$20 per
query.   This system has served this scientific community well, enabling Nobel Prize Winner Selye to write a number of excellent
books with great facility.   This completely manual system stands as
a challenge for those who would improve bibliographic control and
service by automation.

But automation does hold promise in this direction, provided
that objectives of all tasks to be automated and all steps on the way
are carefully and realistically specified.   The reason for optimism
is that some very impressive technological developments are now in
the making, not so much in the area of computers, but in storage
and input-output, and these will broaden the scope of what can be
realistically automated.

4.   Impressive Technological Strides Can Be Expected

I will now briefly sketch some of these technological developments and close with a speculation about the kind of total system
that might be anticipated.

At first,[8] computer technology developments were marked by
the achievement of very high processing speeds.   The current phase
appears to be marked by the achievement of very high capacity, ra-
pid-access storage devices.   The next phase may well be in the attainment of input-output terminals, coupled into large communication
and computer networks, these terminals being as ubiquitous and convenient as the telephone.

In the area of large digital stores, commercially available
magnetic disc files can store close to a billion characters with random access to an addressed record in less than 1/4 of a second.
Also, commercially available tapes can store up to 30 million characters with a serial read rate of 170,000 characters/second.[9]   Advances in newer technologies like those based on flexible discs have

led to expectations that trillion-bit files with random access less than a second could be developed in the near future. This would be adequate for storing the future contents of the National Union catalog. This could make it possible to search for books with certain combinations of subject headings, an example of a task that would be quite difficult if the catalog were in book or card form.

Non-digital image stores are demonstrating impressive size reductions and resulting speeds in retrieval. Linear reductions of 200:1 with good reproducibility have been demonstrated in the laboratory. Small-scale experimental systems providing microimage throughput of better than one per second have also been demonstrated.

In the area of on-demand communications via reactive input-output terminals, there have also been considerable technological advances that merit attention here. Modifications of reservation systems, like SABRE, which can also be used for various scheduling operations might also be usable for on-demand access to vast information files. Reservation systems now under development can link over 100 cities on six continents, handling some 80,000 messages/day. Expected net savings by airlines have been estimated at $2 million/year.

A significant non-commercial time-shared national computer-terminal net now under development is based on the Q32 computer at SDC in Santa Monica.[10] The computer is unusual in that it has 65,000 words of core memory in addition to 16,000 words of special core storage reserved for input-output communication. It is about 2-4 times faster than a 7090 system. Up to 50 teletype stations, 25 of which would be located throughout the country, are planned. Teletype links between Berkeley, Boston, and Washington have been demonstrated. Programs are run in multiples of 50 milliseconds, and the maximum waiting time at any terminal is to be 30 seconds. The Q32 will shortly have IPL V on it, which is significant because this is a convenient language in which to write many experimental programs of interest in information retrieval.[11]

## 5.   What Total System Can Be Realistically Anticipated

Let me conclude by describing a future system that could be realistically envisaged based on what I have said.  In the first place, it is necessary to emphasize that the purpose of this system is to enable the first-rate professional who synthesizes knowledge in his field or who needs it for planning and alerting, to do the work of several men.  The automated aspect of this man-machine team leaves evaluative and motivational judgments to the man.  Its aid is confined to high-speed lookup from large information files, in high-speed routine deduction and possibly in the display of numerous combinations of data suggestive of new hypotheses and queries.

Imagine a scientist or research manager at a console with a very tentative proposal for research.  He wishes to either pursue his idea to the point of a decision to abandon or to implement the project.  Having instructed the system that he wishes to explore a research plan based on an idea, a multiple-choice display appears in response.  This has standard questions with a number of standard categories which apply to most proposals at an early stage. Whenever none of the alternatives are suitable for a particular user, he keys in his own response.  In this way, the scientist provides information about his idea as well as entering a prose description. Any aspects of his entry which can be processed automatically--for instance, checking that the machine's instructions were followed-- may result in immediate feedback in the form of further prestored instructions and questions to the user.  The final proposal is routed and displayed, on request or according to a schedule, at other consoles for people who will evaluate it.  Their aim is to recognize and encourage research based on good ideas, and only those.  To do this, they must not only bring to bear their own knowledge and judgment, but by skillful use of the encyclopedic aspects of this system, they must also have access to all the relevant existing knowledge to support their judgment.  The reserach planner, of course, also has access to the automated encyclopedia.  He may often be greatly helped by being shown just one central idea, fact, or method that he may not even have known how to ask for--provided that he can be led to that item directly.

The evaluators originate a message to be transmitted back to the planner in the same way that his message was routed to them. Several such passes may occur before a final decision is agreed on.

If a final approved plan emerges from this process, the scientist would then like to make sure that someone else will not shortly beat him to the project. This might even be someone who "eavesdropped" on the procedure by which the plan was worked out. There could be thousands of projects any one of which might be competing for the fruits of his labors with a head start. Each of these thousand projects submit reports or show other signs of activity. If one of these projects is knowingly competitive, its leader will try to conceal what the competitor most wants to know. But he cannot, hard as he might try, avoid leaving some detectable traces.

In this kind of detective work, the planner should be able to get help from the same system. The machine could help deduce far reaching consequences from premises supplied by the user. It should be able to combine significant evidence into patterns so as to amplify the detective's ability to guess as well as reason. This would be done by showing the user a sample from a great variety of patterns, excluding obviously useless ones, on the assumption that he could recognize a revealing pattern when he sees it, though he may never have hit on such a pattern unaided. Hopefully, such machine assistance does not disturb his thinking of potentially significant hypotheses he might have discovered without such aid.

To sum up, the message I would like to get across is this. There is a real problem facing management of professional activities due to overloads on the information system, and this problem challenges inquiry and requires contributions from the professionals themselves to integrate knowledge at least as fast as it is generated; from mathematicians and social scientists to devise methods to help them; from engineers to provide the means. I have tried to sketch the nature of this problem and to suggest areas in which progress can be made.

## Notes

1.    Preparation for this lecture was supported, in part, under Contract AF 19(628)-2752.

2.    Status Report on Science & Technical Information in the Federal Govt. 6/18/63-COSI of FCST.   Science, Government & Information, 1/10/63.   P-SAC, Panel on Scientific Info., A. Weinberg, Chairman.

3.    All these figures are obtained from various publications. Neither the authenticity of these figures nor the reliability of their source has been checked.   They should therefore not be quoted without such careful verification.   While it is somewhat justifiable to mention them in a speech to indicate orders of magnitude, this should not be done in the documented literature to avoid perpetuating errors.

4.    On the whole, these would not be considered ''documents'' in the sense in which this term has been defined in this report.

5.    An alternate view of the role of automation is not to suppose the best human capabilities will be amplified and lesser capabilities replaced, but to assume that automation brings marginal people--those of lesser capabilities--back into the labor market to service the machines.

6.    Of course, I could never understand, from any point of view, why right-margin justification is so very important.

7.    This is due to W. Uttal, now at the University of Michigan.

8.    The first phase was, roughly 1940-1960; the second, perhaps 1960-1970; the third, beyond 1970.

9.    10 megabit serial read rates seem to be the state of the art at the time this footnote was added.

10.    This system, together with project MAC at MIT and the Caltech System, represents the state of the art, though the latter is, perhaps, more advanced.   Certainly, the projected effort on MAC is at the frontier.

11.    This is why it is of special interest in connection with this report.

I C.   Toward Information Systems Science: Information Flow
Patterns and Self-Regulating Mechanisms in the Natural
Settings of Libraries[1]

M. Kochen

Abstract:  In search of a scientific discipline underlying the analysis
(and design) of information systems, it appears fruitful to view the
information system of an organism as having three major functions:
1)  to form a parsimonious, coherent, internal picture of its environment useful for guiding action and prediction; 2)  to form an external representation for documenting, storing, and communicating
records; 3)  to improve the internal picture and the store of records
based on information from the environment as well as recorded data.

<u>Introduction</u>

The purpose of automating certain tasks performed by white-
collar and professional labor in an information system is, as in the
case of blue-collar labor, to increase productivity per worker, to
decrease unit cost-to-performance ratio.  A prerequisite for successful automation, therefore, is to understand the precise functions
(tasks) that need to be performed in an information system.

Information systems are found in institutions, in individual living organisms, in entire cultural communities and, presumably, the
same principles may be underlying the functioning of the information
system of all these.  The study of such principles may in several
decades emerge into a new scientific discipline from current concepts and methods in theory of automata, computer programming
techniques, and communication theory and certain aspects of behavioral science (sociology, cognitive theory).

41

## Functions of an Information System

It seems fruitful to view the information system of an organism to perform three basic types of tasks.   1)  Maintenance and use of an internal image of relevant aspects of the natural environment of the organism for purposes of guiding behavior, 2)  Maintenance and use of a store of records for purposes of externally representing internal images to facilitate social communication,  3)  Building and improving both the internal image and the external store of records based on information from the environment and from the store of records.

In concrete terms, an institution, like Harvard, is here viewed as a living organism.   The three functions of its information system may be embodied in 1)  the total understanding within the entire Harvard intellectual community,  2)  the Harvard library system, 3)  the actual community of scholars affiliated with the university.

To get at the principles of such an information system,  it is useful to study the nine interactions among the three functional components.   The units in terms of which component 1 forms internal images of its world are something like elementary symbolic images, perceptions of truths,  simple representations of concepts.   The units in terms of which component 2 stores records are documents,  and the units in terms of which component 3 operates on 1 and 2 to generate improved models and richer document stores are,  for the case of an institution,  people.

Component 2 interacts with itself in that its units are interconnected through document-document coupling,  primarily citations. Study of this type of interaction was recently started (2).

Viewing documents as cultural artifacts,  authors are regarded as agents.   These agents are units of component 3,  so that one bond from 3 to 2 can be read as "is author of."  These agents produce new documents by a synthesis of existing documents mixed with their understanding.   There is thus also a bond from 2 to 3,  namely, "is read by."  Studies on readership and authorship are also recent (1) and there is still relatively little factual data or theoretical insight available.

Agents function by interaction with one another, coupling component 3 with itself. The nature of the bond is something like "influences," in the sense that an author is influenced in what he reads (and writes) by the particular mix of professors, colleagues, etc. with whom he happens to have come together during his intellectual development.

Components 1 and 3 are linked in that a unit of 3, an agent, understands or discovers an element of component 1. A unit of component 1--a concept, an idea, an insight--is, in turn, significant for an agent.

The units of component 1 are linked to each other through associational and logical bonds. Relatively little is known about the four last-mentioned interactions, although there is currently increasing interest from a variety of quarters: "Artificial Intelligence," a revival of S-O-R psychology, the sociology of research and discovery.

The two remaining interactions between components 1 and 2 are somewhat more familiar, though at a superficial level. The units of 2 (documents) are coupled toward the units of 1 (concepts, ideas, topics, truths) by the relation "pertains to." The coupling in the other direction is expressed by the relation "is a theme worth documenting and recording in."

A compilation of some facts concerning several of these interactions is presented in Ref. (3).

## Information Flow Patterns and Self-Regulating Mechanisms

To understand an information system is to understand the dynamics of information flows and processes. There is no natural starting point for describing the dynamics. Data from the environment and from the library are sensed by the agents of component 3. The agents assimilate these data into their collective picture of the world as it appears at the time. They revise this picture as necessary. The new picture helps the agents to screen out the more relevant from all the data and documents to which they are exposed. This, in turn, affects the kinds of new documents of more or less lasting value that will be created for communication to other agents and other institutions.

An agent is a person with finite capacities for interacting with other people, for understanding, for discovery, for reading and producing literature. As total population, and the number of professionals in particular, increases, so does the literature. Hence the fraction of others with whom a given person can be in influential professional interaction decreases. So does the fraction of the literature any one person can know. If the network of people as interlinked through significant professional associations were completely homogeneous--i. e., if the set of colleagues for any person were a random sample from the collection of all people in the institution-- then the likelihood of significant interactions is low and would decrease with size.

Professionals are, however, stratified, and associations within a stratum are more likely than across strata. Consequently, the population of an institution fragments into slightly overlapping professional cliques. For each clique there is a communality of understanding, a characteristic portion of the world picture on which they are expert. There is also a corresponding cluster of documents unique for each clique. The match between such a document cluster and the appropriate clique is effected by the evolution of a suitably specialized jargon.

Can automation significantly increase the productivity of a professional in component 3 of an information system? Suppose that a professional could productively interact, and be significantly influenced by at most 100 others at any one time on a particular topic; possibly, automation could assist in helping him select and encounter a more suitable set of 100 from all the possible encounters, than the 100 to which a person is normally exposed. This is not to suggest the elimination of chance encounters; on the contrary, the selection of the optimal set of 100 may well use a chance mechanism. Similarly, if a person can only spend 8 hours a week reading the literature, perhaps automation could assist in helping him select the right (increasingly small) fraction of all the literature.

As the use of bibliographic tools like citation or, permuted-title indexes becomes widespread, authors' practices of titling and referencing will be profoundly affected. An author, knowing that his paper is likely to go undetected if he does not accurately acknowledge intellectual debts to prior documents, will be under strong motivation to do so. These are but a few of the more elementary kinds of regulatory mechanisms that are now understood.

## Conclusion

The populations of all three components of an information system are growing, possibly at the same rate. That is, the cumulated number of documents is doubling every 8-15 years; the number of professionals is also presently increasing at an exponential rate of about 7% a year; how much is known[2] is also growing rapidly, although there is as yet no good way of measuring this. Presumably, however, the latter is also growing, as fast as ever, and this growth would bear witness to the effectiveness rather than to pathologies in our present information system, as so many who view the literature explosion with alarm would imply.

The mechanisms whereby these three growing entities remain in stable equilibrium are those of fragmentation (classification, clustering, grouping) and integration (hierarchic and other ordering relations and processes). As population continues to expand, and technology makes further advances in speeds of communication and information processes, it becomes increasingly critical that the regulatory mechanisms continue to ensure stability. Safeguards against possible instability could be designed into information systems with the same technologies of high-speed and high-capacity communication, storage and processing that contributed to the danger of instabilities in the first place.

## Notes

1. This work was supported in part by Air Force Contract

AF19(628)-2752.

2.  The unique and curious thing about knowledge in this context is that, as it grows, so does our awareness of what is not
known.  It is perhaps true that for every new truth that is discovered, several puzzles on a related topic that were not recognized
before become clear and may lead to new truths.

### References

Bureau of Applied Social Research, Columbia University, Rev.
of Studies on the Flow of Information Among Scientists, January,
1960, H. Menzel, Study Director.

Kessler, M. M., "Bibliographic Coupling Between Scientific
Papers," Proc. WJCC, Vol. 19, May 9-11, 1961.

Kochen, M., "On Natural Information Systems: Pragmatic Aspects of Information Retrieval," Methods of Information in Medicine,
Vol. 2, No. 4, October, 1963, pp. 143-147.

# I D.  Preliminary Operational Analysis of a Computer-Based, On-Demand Document Retrieval System Using Coordinate Indexing

M. Kochen

## Introduction

The purpose of this paper is to illustrate what needs to be done and what further facts need to be established prior to the thorough quantitative analysis of a document retrieval system. The reason for such an analysis is to determine specifications of required hardware and software, to compute performance prior to costly engineering design and construction, and to estimate economic factors. I do not particularly endorse the coordinate indexing method on which this analysis is based. I chose it because it facilitates the systems analysis, corresponds to existing practices, and because it is pedagogically useful.

## Model for the System

The system to be analyzed consists of the following equipment:

1) a central store and its associated search logic unit
2) a communication net connecting the store, through buffers, to on-line terminals
3) a central processor allowing access to the store from the controls in a time-sharing or multiprogramming mode and capable of matching pairs of numbers or words.
4) terminals allowing on-demand querying for catalog records of documents in a large collection.

The store contains a table. Its arguments are an unordered list of coded descriptors or main entries. Corresponding to each argument, the tabular entry is a list of pertinent document identifi-

47

cation tags.   There is a second table, which stores for each document identification tag a brief description of the document, suitable for display.

To use the system, a querist enters a disjunction of conjunctions of descriptors, at a terminal; for example: (carcinogenic & chromium) $\not/$ (cancer-producing & chromium) $\not/$ (carcinogenic & Cr) $\not/$ (cancer-producing & Cr).   After searching for all such query terms among the arguments of the stored table, a list of document identification tags for each term is read out.   The tags common to all these lists constitute the required bibliography; the corresponding document descriptions are displayed.

All this is very conventional and simple.   It is not difficult to suggest a more sophisticated system, and it is toward the analysis of such a system (AMNIPS)[Ref. 1] that this analysis is a first step.   The detailed analysis of even this simplest system reveals difficult and interesting problems.

### Basic System Parameters

These variables are defined here in summary fashion, and their order of magnitude for many applications is given in parentheses.

D = total number of documents.  $(10^5)$ Note 2

d = average number of descriptors (index terms)/document  (10, range: 1-100)

T = total number of different descriptors  $(10^4)$

t = average number of documents posted on a descriptor  (100, range: $10-10^3$)

D' = rate of growth of collection, in no./month  $(10^3)$ Note 3

W = maximum tolerable waiting time, in seconds, from acknowledgment to beginning of printout of a bibliography, in response to a query at a terminal  (5-100, for on-demand) Note 4

$N$ = number of terminals connected to
central processor (20)

$U$ = total number of subscriber-users (100)

$u$ = average number of queries/month/
user (100) Note 5

$h$ = hit-rate or ratio of expected number
of documents both retrieved and relevant to expected number of relevant
documents (. 3-. 8)

$a$ = acceptance-rate, or ratio of
expected number of documents both
retrieved and relevant to expected
number of retrieved documents (. 1-. 9)

$M$ = total storage capacity of central
files in bits $(10^9)$

$T$ = average access time, in milliseconds, to locate a record in the
auxiliary store, given a descriptor
which acts as keyword to the
record (100)

$s$ = serial read rate, in bits/second $(10^4-10^7)$

$T'$ = the cycle time, in microseconds,
of the central processor (2)

$m$ = average number of different terms
per query (put in conjunction) (3)

$n$ = average number of near-synonyms
per query term (4)

$c_c$ = cost of the central files and processor ($/month) (50, 000)

$c_t$ = cost of a terminal ($/month) (500)

$c_I$ = cost/document of adding a new document record to the file in $

$p$ = price/query a user is willing to pay,
in $ (10)

The number of second/month is assumed to be $5 \times 10^5$.

The basic figure of merit is total net profit, P (\$/month), and the above variables are to be chosen so as to maximize P, subject to several constraints described in the next section.    Profit and price is chosen here only to allow comparison of what is gained and given up in using such a system.

It is assumed that this system will be used only for document retrieval and not for computation, payroll, or other numerical operations.    This assumption is antithetical to the idea of time-sharing.

## Assumptions and Basic Relations

1.    It is easy to see (and well-known; see Wooster & Taube)[Ref. 2] that the total number of term-items is

$$Dd = tT \qquad \text{Eq. (1)}$$

This equation also holds when each of the four variables in it are replaced by means[6] of the corresponding random variables in a more refined analysis.

2.    Telephone congestion theory suggests that the number of simultaneous users (at peak load) is approximately 1/5 of the total number of subscribers.    Thus

$$N = (1/5)U \qquad \text{Eq. (2)}$$

This means that five subscribers are to share a terminal.[7] This is, of course, a safe upper bound on N to avoid queuing at terminals. In a more defined analysis, the distribution of interarrival time, service time, etc. needs to be checked against the assumptions of telephone congestion theory.    (See Syski)[Ref. 3]

3.    Suppose that each query is posed in terms of a disjunction of clauses; each clause consists of m nonsynonymous terms, and all the n near-synonyms for each word are used.    Thus, there will be $n^m$ disjunctive terms.    It can be shown that the mean time to retrieve the list of documents indexed by all the prescribed terms is

$$mn(r\ 10^{-3} + s^{-1}t\ \log_2 D)$$

$$+ \quad 20 r' n^m t^2 (1 - \frac{t}{2D})\ \frac{(1 - (\frac{t}{D})^{m-1})}{(1 - t/D)}\ 10^{-6}\ \text{sec.} \qquad \text{Eq. (3)}$$

The first term is the time required to read mn lists, from the auxiliary store holding the index file, into core. There is a list for each of the mn terms in the query. Each list consists of t document numbers, on the average. Each document number must be at least $\log_2 D$ bits long.

The second term is due to an approximation, derived by E. Wong,[Ref. 4] and is based on assumptions of statistical independence among query terms. It holds for the case in which all the lists of document tags are unordered. The dominant term is $t^2$, which estimates the number of comparisons to determine the documents common to the m lists for the terms of one clause in the query. The average time to make one comparison is the time it takes to

   a) transfer an element of a list from a core address to the arithmetic unit

   b) subtract an element from another list

   c) check if the two match, and transfer control to

   $c_1$) place the matched element on a new list in case of match, or

   $c_2$) repeat steps a–c for another pair of elements in case of no match.

Assume that step (a) takes about 2 cycles, step (b) about 2 cycles and step (c) about 2 cycles. Assume further that step ($c_1$), involving an add, store and address updating instruction, takes about 10 cycles, that $c_2$, involving an address-updating instruction, a check to determine if all pairs of elements have been checked, and a transfer, also takes about 18 cycles. The average time of steps $c_1$ and $c_2$ [with a drastic assumption of an equal chance for the two outcomes of decision (c)] is about 14 cycle times. Thus, the total time to compare is about $(6+14) \tau^1$ microseconds.

With the reasonable figures used later, this processing time/ query is about 25 seconds.[8] The time to read the descriptive catalog records corresponding to all the documents that satisfy the specifications of the query is also neglected in this analysis. So is the time the terminal is kept busy printing. If done off-line, buffers are required, and these are expensive. All these considerations should either enter into a revised form of Eq. (3) or into the cost of a terminal in a future phase of this analysis.

Eq. (3) expresses the processing time/query, and it should be equal to $\frac{W}{N}$ if the N querists who are on simultaneously at peak load are each to be served in W seconds. This, too, is an upper bound, assuming that a user will have to wait for all others to have their turns before he gets serviced. Thus, approximately,

$$W = N \left[ \quad mn(.001\ T + \frac{t}{s} \log_2 D) \right.$$

$$\left. + 20T'n^m t^2 (1-\frac{\theta}{2}) \frac{1-\theta^{m-1}}{1-\theta} 10^{-6} \right] \quad \text{where } \theta = t/D \qquad \text{Eq. (4)}$$

4. The simplest[9] way of assigning descriptors to documents is to use all the different word-types in the document except the very common and the very rare ones. (In this case, d would be of the order of 100 rather than 10, and t would be 10 times as large also.) There will be as many entries to be updated into the central files as there are word types. The time to check for the recurrence of word-types within the document will be neglected. Since the descriptor files are assumed unordered, the average number of comparisons to be made for each word-type will be $T/2$, unless novel key-address transformations are used. Hence, it will take about $d(\frac{T}{2}[\frac{30}{s} + 20T'\ 10^{-6}] + \frac{\log_2 D}{s})$ seconds/document to update the file. This assumes about 30 bits/index term. Since the file grows at $D' \times \frac{1}{5} \times 10^{-5}$ documents per second, the fraction of the time available for updating which is required is about

$$d \left( \frac{T}{2} \left[ \frac{30}{s} + 20T'\ 10^{-6} \right] + \frac{\log_2 D}{s} \right) \frac{D'}{5}\ 10^{-5}.$$

In one second, in addition to processing $\frac{1}{5}\ 10^{-5} D'$ documents for file updating, it is necessary to process on the average, $\frac{1}{5}\ 10^{-5} uU$ queries. Thus, it is necessary that $\frac{dD'}{5}\ 10^{-5} ( \frac{T}{2} [\frac{30}{s}$

$$+ 20\ T'\ 10^{-6} ] + \frac{\log_2 D}{s} ) + \frac{u}{5}\ 10^{-5}\ \frac{W}{N} \leq 1 \qquad \text{or}$$

$$dD' \ ( \ \textsf{T} \neq s^{-1} \log_2 D \ ) \ \neq uW/N \ \leq 5 \times 10^5,$$

$$\text{where} \ \ \textsf{T} = \frac{T}{2} \ ( \ \frac{30}{s} \neq 20 \ \textsf{T}' \ 10^{-6})$$   Eq. (5)

For 100% utilization (without queuing) of the central processor and file, the equality sign in Eq. (5) should hold, but then Eq. (4) may not hold for a prescribed W. [Note 10]

If the terminals are to be used for entering new documents, a similar relation must hold for each terminal.

5.  A critical variable to be determined is U, the number of users.  The simplest demand curve for U is

$$U = U_0 \ 1/p$$   Eq. (6)

where p is the price a user is willing to pay.

But p is a function of the quality of retrieval.  Essentially quality can be measured by three parameters: hit-rate, acceptance-rate and waiting time.  It is a reasonable guess--to be replaced by one based on facts when these are available--that p decreases as a power function with W for W exceeding a certain minimum (milliseconds), say as $W^{-b}$.  The higher the hit-rate and acceptance rate, the more a user should be willing to pay.  The simplest, though probably unrealistic assumption is proportionality to the product ah. Thus,

$$p = BW^{-b} \ ha$$   Eq. (7)

Of course $p = BW^{-b} \ (\propto h \neq \beta a )$ would be another, equally plausible choice, one to be explored no further in this paper.

6.  Next, it is necessary to estimate the relation between a, h and the basic system parameters m, n, d, T, t and D.  A new basic variable must also be introduced.  It is assumed that the set of all index terms is interlinked in a graph which is organized into T/L "clusters" of L members each.  The terms in one "cluster" are more strongly related--synonymy is one such relation--each to every other member of the cluster than to members outside it. [11]

A document is said to be relevant to a single query term if its index set--the set of d descriptors assigned to it--contains at least one member of the cluster to which it belongs.  Each of the d index terms is assumed to be part of a different cluster. Let c be the conditional probability that the set of d index terms contains one of the $n^m$ clauses of the query, given that the m clusters represented by the query are present among the d clusters represented by the index terms. Each cluster is represented by n terms in the query, and the chance that the representative of the same cluster among the index terms is one of these n is n out of L for each cluster. Assuming the different clusters in a query to be independent, we have

$$c = (n/L)^m \qquad\qquad \text{Eq. (8)}$$

Consider now the corresponding conditional probability c', that the set of d index terms contains one of the $n^m$ clauses of the query, given that none of the m clusters represented by the query are present among the d clusters represented by the index terms. We assume that a query term can (erroneously) be among the index term because it is a homograph or just because it is a common term which occurs in the text without making special reference to its topic; we suppose that the chance of one of the T terms being among the d index terms is just d out of T, and since there are n near-synonyms for each word, the probability of a word or one of its near-synonyms being among the d index terms by chance alone is, approximately, dn/T. The probability c' is given by

$$c' = \binom{d}{m} \left(\frac{dn}{T}\right)^m \left(1 - \frac{dn}{T}\right)^m \qquad\qquad \text{Eq. (9)}$$

It is known[Ref. 5] that, under general conditions, the hit-rate h, is equal to c and the acceptance rate a is equal to $c\theta \,/\, (c\theta + c'(1-\theta))$, where $\theta$ is the probability of a randomly chosen document being relevant; this can be estimated by $\theta - d/T$. Thus, approximately,

$$c' = \binom{d}{m} (\theta n)^m \, e^{-mn\theta} \qquad\qquad \text{Eq. (10)}$$

Hence,

$$a = \frac{\theta}{\theta + (1-(n/L)^m) \, (L\theta)^m e^{-mn\theta}}, \quad \text{with } \theta = d/T - t/D$$
$$\text{Eq. (11)}$$

$$h = (n/L)^m$$

7.    To compute the total required storage capacity of the central file, consider that there are T tabular entries. Each of these T arguments requires at least $\log_2$ T bits, but this minimum is achieved at a cost in lookup time; an estimate of 60 bits/argument is more reasonable. The function corresponding to each argument

consists of t document numbers, each at least $\log_2$ D bits long. For display purposes there must also be, say, k bits of information about each document in a separate store.

Thus,      $M = T(t \log_2 D + 60) + kD$                Eq. (12)

8.   The remaining problem is to interrelate the various costs. For the cost of the central processor and files, a first approximation might be

$$c_c = C_c \frac{M}{T}$$                Eq. (13)

The cost of the terminals could be assumed to decrease with their number, as $c_t = C_t (C'_t + e^{-N})$.   For present purposes, however, $c_t$ is assumed constant.

The cost of entering a new document into the file depends on whether the central file is ordered or not.   A random access file should be ordered.   This reduces mean search time but increases updating time because of merging operations.   This is significant if the activity of the file (u) is comparable to its growth (D').   Updating in batches may help.   An unordered serial file sacrifices mean search time for ease of updating.   This is the case that has been studied here, as a start.

In computing $c_I$, the computer time for indexing must not be counted again, provided there is enough computer time to do all the selection of words, merging, etc.   If the same terminals used for querying could also be used for converting incoming documents to machine-readable form, the only extra cost is that of conversion labor, about 1 cent/word with presently available key-punching and verifying techniques.   Assuming the average document to be 1000 words long, $c_I = \$10D'/$month.                Eq. (14)

Maximization of Profit

With the (reasonable) assumptions made so far, it is difficult to find combinations of parameter values which lead to any profit at all, let alone maximization of profit.   The basic equation for profit, in \$/month, is the total income less the total cost:

$P = Uup - c_c - Nc_t - c_I,$

This may be written, using Eqs. (13), (2) and (14), as:

$$P = Uup - C_o \frac{M}{T} - \frac{1}{5} Uc_t - 10D'$$          Eq. (15)

From (7), (8) and (11), we have

$$p = B \left[ W^{-b} \right] \left[ (n/L)^m \right] \left[ \frac{\theta}{\theta + (1 - (n/L)^m) (L\theta)^m e^{-mn\theta}} \right]$$

Note that the last bracketted term, the acceptance rate, is above
what would be obtained by chance--the value $\theta$-- $L < \frac{1}{\theta} \sqrt{1-\theta}$

or $L < T/d$. We choose some illustrative numbers, taking L slightly
larger than n=3. Reasonable values of L, d, T, and m which satisfy this condition are: d = 25, $T = 10^4$, m = 2, L = 3.22, so that
$\theta = .0025$. This makes the acceptance rate approximately .131.

To avoid obscuring fundamental issues with lengthy formulas,
it is useful to substitute sample additional numerical values at this
point. Taking n = 3, the middle bracketted term, which is the hit
rate, is about .86. From Eq. (1), and the value $D = 10^5$, it follows that $t = \frac{25 \times 10^5}{10^4} = 250$.

To compute the waiting time W according to Eq. (3), take
$s = 10^4$ bits/sec and $T' = 2$ microseconds. Then
$W = N [ 6 (.001 T \neq .425) \neq 22.5 ] = \frac{1}{5} U (25 \neq .006 T) = (5 \neq .0012 T) U$.

Taking B = 100, the price/query is about
$$p = 11.6 [ (5 \neq .0012 T) U ]^{-b}$$          Eq. (16)

From a demand curve like that of Eq. (6), and Eq. (16), it
follows that
$$U = \left[ \frac{U_o}{11.6} (5 \neq .0012 T)^b \right]^{\frac{1}{1-b}}$$          Eq. (17)

This suggests making T the critical independent variable, as
indeed it should to estimate the advantages due to automation; b is

a critical parameter; the nature of the demand curve, however, is also critical.

To further simplify equation (15), take $u = 10$, $C_0 = 10^{-3}$, $c_t = 500$, $D' = 10^3$, $k = 10^3$. Then $M = 10^4 (4310) + 10^3 \times 10^5 = 1.43 \times 10^8$ bits, and

$$P = 10 \ U_0 - 1.43 \ 10^5 \ T^{-1} -$$

$$100[ \frac{U_0}{11.6} (5 + .0012 \ T)^b ]^{\frac{1}{1-b}} - 10^4 \qquad \text{Eq. (18)}$$

The system can be profitable, i.e. $P > 0$, if $T > 1$ and if b is close to 0. If $b = 0$ users would pay no more for on-demand service than they would for service with arbitrarily long delays, about \$11.6/query. In that case, taking $T = 100$ milliseconds,

$$P = 10 \ U_0 - 1.431 \times 10^3 - 8.6 \ U_0 - 10^4 = 1.4 \ U_0 - 11,431.$$

Profit results if $U_0 > 166$, the number of subscribers that would buy the service at \$1/query, exceeds 8200.

Let us, however, examine if this is feasible, by checking if Eq. (5) (top equation) is satisfied. Note first, that the waiting time/query is 5.12 U seconds; for $U_0 = 8200$, $U = \frac{8200}{11.6} = 707$, so that $W = 60$ minutes! Substitution into Eq. (5), gives $\frac{10}{5} \times 10^{-5} \times 25.6 +$

$$+ \frac{25 \times 10^3}{5} \times 10^{-5} ( \frac{10^4}{2} [ \frac{30}{10^4} + \frac{40}{10^6} ] + \frac{17}{10^4} ) = 5 \times 10^{-2} (15 + .2 + .0017) +$$

$+ .0005 = .76.$ This is feasible, with 76% computer utilization.

Suppose, however, $b = .5$. Then,

$$U = \frac{U_0}{11.6} \sqrt{(5 + .0012 \ T) \ U}, \text{ and}$$

$$P = 10 \ U_0 - 1.43 . 10^5 \ T^{-1} - \frac{U_0^2}{135} (5 + .0012 \ T) - 10^4 \qquad \text{Eq. (19)}$$

No values of $U_0$ and $T$ will make $P > o$ in this situation.

## What Needs to Be Known

In order to make a thorough analysis, the results of which can be relied on for planning technological developments and marketing

information systems of this sort, [12] the following should be established, probably by empirical or experimental methods.

(1)  A demand curve, such as Eq. (6)

(2)  A price-quality curve, such as Eq. (7)

(3)  Cost relations, like Eqs. (13), (14).  Certainly, in Eq.
(13), the cost of the central depends not only on $T$ but also on $T'$ and s.

Other relations also need further development.

(1)  More adequate measures of performance than the hit-rate
and acceptance-rates are necessary, and these must be expressed in
terms of the system parameters.

(2)  Expressions are needed for the waiting time in ordered
files, analogous to that in Eq. (4); also expressions relating $T$, $T'$
and s.

(3)  Expressions for total storage capacity in more sophisticated index files than the ones used here, to replace Eq. (12).

Additional constraints exist.  For example, teleprocessing
may be most economical if confined to within a 10-mile radius or
so, and using twisted-pair transmission lines.  This sets an upper
limit to the number of users according to geographical constraints.

Nearly all the variables used here should be treated as random
variables.  Many of the independence assumptions must be reexam-
ined.  Perhaps the variable u should depend on p, up to an asymp-
tote.  (As the cost per toll unit goes down, does a telephone subscriber make more calls/month?).  The entire analysis needs to be
repeated for sophisticated retrieval systems.  The AMNIP system,
for example, permits one to investigate how the use of a citation
index affects h, a, W; how the use of a thesaurus affects these variables; how the use of predicates in addition to only index terms affects them.

One of the important later steps is to introduce time to model
the dynamics of the situation.  The expression for the price/query
users are willing to pay depends, in part, on how much it would
cost to supply the same service without automation.  These costs go
up not only because $D'$ increases geometrically with time, but be-

cause labor costs are increasing. The costs $c_c$, $c_T$, $c_T'$, on the other hand, could be expected to decrease with time.

Concluding Remark

The model sketched here, and its use for calculating when a time-sharing reference retrieval system provides useful service relative to its cost, is meant to illustrate the kind of "systems arithmetic" that could and should be made whenever a retrieval system is proposed.

Notes

1. This work was partly supported under contract AF19(628)-2752.

2. This is what we regard as a critical or minimal-size date base for significant document-retrieval experiments.

3. A 7 per cent / year geometric growth rate might be reasonable for younger, specialized information centers. A lower figure would hold for older, established, comprehensive libraries.

4. The printing of the bibliography would be done off-line. Perhaps 25 per cent of the queries may need on-demand service; the remainder might accept 24-hour service.

5. This figure may be high, but varies greatly with the subject matter and type of querist; $u = 1$ is not unreasonable either.

6. This is an interesting ambiguity. The reader will surely know from the context what is meant by "means"!

7. This assumption, and the corresponding loss of convenience, may enter into the relation for the price per query a subscriber is willing to pay, Eq. (7). One of the great requirements of an on-demand system is that its terminals must be as convenient, accessible and ubiquitous as telephones.

8. At 10 cents/sec. (i.e. $50,000/month), this would cost $2.50/query.

9. Because of the large number of documents (i.e. 250) retrieved in response to a query for such "deep" indexing, this could well be as foolish as it is simple. Yet it is being done.

10.   In Eq. (5), it is assumed that the extraction of the d descriptors from the full text is done off-line or manually.   If the
central processor is used for this, the fraction of time it takes up
must be added to the left-hand side.

11.   In a more sophisticated analysis, these clusters do not
have sharp boundaries, and they are somewhat overlapping.   Relations among names, other than dyadic ones, must also be taken into account.

12.   This is actually a secondary intent of this analysis.   The
first aim is to attempt to compare proposed systems before they are
"tried out" at great cost to eliminate obviously absurd ones, and to
provide a framework for evaluating those that have been built.

## References

Ref. 1.   Kochen, M.   See IA of this volume.

Ref. 2.   Wooster, H. and Taube, M. ed., Information Storage and
Retrieval Theory, Systems, and Devices, Columbia Univ.
Press, New York, 1958.

Ref. 3.   Syski, R., Introduction to Congestion Theory in Telephone
Systems, Oliver and Boyd, Edinburgh, 1960.

Ref. 4.   Wong, E., "Time Estimation in Boolean Index Searching"
in High-Speed Document Perusal, final report by M. Kochen
to AFOSR under contract AF49(638)-1062, May 1, 1962, pp.
69-81.

Ref. 5.   Kochen, M.   "Toward Document Retrieval Theory: Rele-
vance-Recall Ratio for Texts Containing One Specified Query
Term," Proc. ADI., Part 3, Chicago, Oct. 6-11, 1963,
p. 439.

## II.   The Knowledge Subsystem

We examine first the subsystem of an information system
which is responsible for representation, comprehension, integration
of accumulated information in more detail.

In modeling information systems, it is useful to consider them
at different levels:

A.   The information system (and its three subsystems) in a
national institution as it evolved over centuries, (e. g.   Congress
with (1) the Library of Congress, (2) the legislative community,
(3) the corpus of law and data pertinent to pending legislation in its
jargon).   This is comparable to:

B.   The information system and its three subsystems in les-
ser-scope institutions as these evolved over decades.   (e. g.   the
Engineers' Joint Council with (1) the Engineering Societies Library,
(2) the engineering community, (3) engineering science and know-how,
in its jargon).   These are also comparable to:

C.   The information system and its three subsystems of an
individual as he developed over years (e. g. a corporation executive
with (1) his personal library and files, (2) his staff and several
''cliques'' of colleagues, (3) his expertise).

To go one step further, these information systems and their
three subsystems may also be comparable to those of an individual
in a particular role (e. g. a doctor faced with choosing a treatment
for a given disease with (1) his own memory and pertinent documents to aid recall, (2) his reasoning and comprehension capacity,
(3) his personal experience and knowledge).

In exploring these possible parallels, we seek to abstract what
is common to (as well as classify according to fundamental differ-

ences), for example, the knowledge of a doctor in treating a specific disease, the expertise of an executive, the body of engineering knowledge, the corpus of U. S. law.   We note that these are all examples of discourse.   But what is a significant way for classifying types of discourse?   One gross way to distinguish between types of discourse is to separate them into cognitive and evaluative.   Each of these two types can be further classified according to whether the substance of discourse is:   1.   direct sense data; 2.   suitably abstracted (i. e.  linguistically or otherwise represented) facts, findings, concepts, ideas, which are internal maps of the real world; 3.   explanatory in nature and deals with abstract variables, relations, principles.   Such a classification is presented in the first appendix to this chapter, II A.   Its significance is in the overall perspective and systematization it provides for analyzing discourse. A good systematization of methods of discourse is prerequisite for the development of deeper and more rigorous theories of knowledge and linguistic performance.   The paper II A correlates many ideas from sociology, psychology, philosophy with corresponding ones in information science, and such correspondences are frequently undetected and often rediscovered under different names and in different disciplines. [1]

One of the first requirements of a theory of knowledge (or cognition) is that it must not be restricted in subject matter.   There can be restrictions on the form but not on the substance of items of knowledge represented, say, linguistically.   Instead, it must be capable of explicating the possibility for any aspect of the real world to be known.   This suggests that the mechanisms of language, as related to the formation of internal (or culturally assimilated) models of the relevant real world, should be studied closely.   The thesis which appeared to us worth pursuing was that an entity capable of comprehension--both ordinary language and the real world--must be "nurtured" or grown into existence, rather than predesigned. To pursue this, a marriage between Uhr's pattern-recognition methods, recently extended to the recognition of linguistic patterns, with some of our methods, resulted in a model for the evolution of "organisms" capable of comprehending organisms which are immensely more complex than themselves. Our approach is described in the second appendix to this chapter, II B, by Kochen and Uhr, which grew out of our collaboration on the study of computer comprehension of ordinary language at RAND in the summer of 1963.

One of the key intellectual problems in pursuing this line of thought involves the idea of complexity. How is the evolution or growth of "organisms" capable of comprehending an environment immensely more complex than they are theoretically possible? To approach this question, both "organism" and its "environment" were modeled as finite-state automata. The "organism" has two tasks: to generate the most complex binary sequences it has capability for at any time; to analyze (predict) binary periodic sequences it receives as input, in terms of the rules by which that input could have been generated. "Complexity" of the organism-automaton refers to the ability to do both tasks.

With our present understanding of finite automata, an automaton must be of great complexity--defining complexity as the number of states--to completely predict the behavior of a simple one. But this is contrary to the situation of interest here. Hence, new concepts of automata theory need to be explored.

S. Winograd began such explorations in a preliminary way as part of our study. He searched for a metric to be attached to the inputs as well as to the outputs of an automaton. The "prediction" capability would then depend on the ability of the automaton to "approximate" (in the sense of the chosen metric) patterns of "significant" behavior. Such an automaton, he argued, should have some "adaptive" capability for measuring complexity. The behavior into which the automaton finally settles should be given greater weight than transient aspects; i. e. than states which are only infrequently traversed.

If an organism-automaton is coupled to another of much greater complexity, can the latter analyze every output generated by the former, and are there sequences generated by the latter that the less complex organism cannot analyze?  Can the less complex organism learn to extract from its input data enough about the structure of its more complex environment so as to form an "internal model" adequate for prediction and effective decision-making?

Some of these questions were examined by H. Yamada as part of this study.  A rigorous formulation in its relation to automata theory and literature search indicated that this was a relatively untouched but important area of automata theory requiring more intensive investigation.

Another approach to this problem area, using list-processing techniques rather than the notions of automata theory and pattern-classification, is discussed in II C.  This theoretical study concentrated more on concept and language learning in young children than on these processes in an abstract organism-automaton.  The partially programmed simulation involves an "organism," which has before it "sense data" and "verbal input" and the power to perform a set of "actions" including verbal output and manipulation of senses. The organism continually guesses at the meanings of new words and at the utilities of various actions under different conditions.  Among its list of actions is one allowing it to add other responses to its action list.  This enables it to grow long association nets representing its guesses about the interpretation of current verbal or sense inputs.  Unlike the general-problem-solver, these programs form their own sets of rules and routines for linguistic analysis as they run.  The technique involved in simulation of concept/language-learning includes many of the behaviorists' observations regarding human actions, such as conditioning, stimulus generalization, "differential reinforcement," etc.; it also goes beyond these, however, in forming structures of routines similar to those called "plans" by Pribram, Miller and Galanter.

Both papers II B and II C aim to demonstrate computer programs which exhibit learning of items the content of which is not a priori specified.  Such programs can furnish the base on which to build optimal learning programs or devices or models to describe learning in real organisms. Learning, in the sense of acquiring or accumulating and assimilating knowledge is, after all, one of the key notions in information science. We are here not concerned with acquisition models for language, the outputs of which are generative grammars. The controversy in linguistic circles whether to make minimal assumptions about the structure of organisms enabling them to acquire the ability to generate sentences or whether the ability to generate grammatical sentences is built in is of secondary concern to us.

At a less abstract level, interrelations among concepts, combinations of concepts into ideas, etc. can be studied in terms of interrelations among words denoting these concepts, combinations of words into sentences, etc. A currently fashionable and simple device for interrelating words as these interrelations reflect conceptual ones is that of a thesaurus. A thesaurus also has purely utilitarian value in its own right, as an aid to authors in finding and choosing words best suited to their interests; more recently a "thesaurus" has also come to be a pragmatic tool to aid catalogers and querists in a document storage and retrieval system. As such, the purpose of a thesaurus is to help effect a match between the terms used by an author in writing, by a cataloger in indexing, or by a querist in searching for a document. There have been many attempts to construct a thesaurus once and for all. Such a thesaurus can soon be obsolete, because of the many changing interrelations among words likely to be useful. A method for continually "growing" and using a thesaurus has been developed in connection with AMNIPS. Such a thesaurus system, if implemented and used, is a practical device for enabling the "knowledge subsystem" to grow, for learning to occur. Paper II D describes this adaptive thesaurus system as well as the more general problems and concepts in the construction of thesauri.

Further results, involving the use of a thesaurus to minimize storage and search time are reported in IV C.

Viewing an AMNIPS thesaurus as a graph in which vertices are words, and in which the edges are near-synonymy relations between them, the applicability of some novel results in the analysis of probabilistic graphs was studied. These results, e. g. expressions for the probability that the random removal of k edges from a completely connected graph with n vertices will result in an unconnected graph, are also of interest in their own right. But in this context, they allow comparisons for consistency in the way two people interlink near-synonymous words in a given set of such words, and tests for significant deviations in their word-associations. These results apply to the design of thesauri, and were derived to illustrate the central role that concepts and theorems of graph theory play in a more rigorous approach to describing various aspects of our information system. These mathematical results are appended to this chapter in paper II E.

## Notes

1. The detection and conceptualization of such parallels may, in spite of, or perhaps even because of, obscurities in the language used to present them stimulate theorists toward the formulation of significant special and later general models for behavior.

## II A.  Systems of Orientation[1]

Richard Jung
Cornell University[2]

### Introduction and overview

This paper describes an attempt to formulate a functional theory of orientation.  It is the second working paper in a projected series.  The first working paper[3] reported that a phenomenological analysis and reconstruction of one phase of the postulated process of orientation yielded a conceptualization of twelve systems of discourse.

The present paper reports the results of a formal analysis of another phase of the process of orientation and its reconstruction as a set of systems of orientation.  It is postulated that the systems of orientation are the generative mechanisms for the systems of discourse.

These results are presented in several Figures and Tables. The Figures locate the phases of the process relative to each other. The Tables display the structure of the systems of orientation, and they indicate which components of each system of discourse are generated by a given set of operators in the corresponding system of orientation.  A systematic study of the Figures and Tables should convey efficiently and adequately the central thesis of the paper.

### The theoretical context

The theoretical context of the problem of orientation is firmly established by the major traditions of psychological and sociological theory.  In most theories of behavior, it is regarded as axiomatic that the organism does not respond to the total environment, but to

some selected subset of it.  It is further assumed that every organism takes an active part in constructing its action-world; it not only
selects from the ontological world, but also adds some elements
and imposes some, if not all, relations.

In sociology, since Weber, a necessary step in the analysis
of action is its "meaningful interpretation" (deutend verstehen) in
terms of, what Thomas called, the actor's "definition of the situation." This mode of sociological analysis has been codified and
further developed by Parsons, and its basic concepts are now generally accepted.

Similarly, in psychology, part of the variation in responses of
an organism to ontologically similar conditions is accounted for by
postulated differences in the organism's response set.  The active
part of the actor in constructing his action-world is emphasized in
the classical Freudian, Gestalt, Lewin's, Piaget's and New-Look-
Cognitive theories.  The learning theory of Tolman, the psychology
of personal constructs of Kelly, Festinger's cognitive dissonance
theory, Orne's methodological critique of psychological experiments
in terms of the subject's construction of a meaningful experimental
situation by his detection or invention of the demand characteristics
of the experiment, and the work of Berlyne on epistemic behavior
are among the more recent formulations which assign critical importance to the actor's construction of his action-world.

The problem of orientation is a familiar one in ethology since
von Uexküll's construction of the concepts Umwelt and Wirkwelt.
Similar examples from theories of economic and political behavior
abound.  On the whole, current theories of behavior attribute a significant portion of the variation in behavior unexplained by them to
orientation as an undefined residual category.

It appears that this is not due to the admitted inadequacy of
these theories, but that even fully adequate theories of motivation
and decision will not be able to account for all the variation in action, and that a substantial portion of this residual variation is indeed attributable to differences in orientation.  At the present, no

theory of orientation exists. The construction of such a theory appears to be a necessary and critical step in the further development of behavioral theory, as well as a prerequisite for the programming of mechanical systems for behavior of the human order of complexity.

The aim of the project is to formulate a theory of orientation, i. e. to give a systematic general description and an explanation of the process by which any behaving system (organism, collectivity, or machine)[4] constructs a definition of itself as an actor and of the world as its situation.

Among the major questions the theory should answer are:
(1) What are the different possible types of orientation?
(2) What is the specific principle governing each type of orientation?
(3) What is the principle governing the selection, or the relative preponderance, of the different types of orientation within the general process?
(4) What is orientation, i. e. what is the common nature of all the different types of orientation?
(5) What is the principle governing (i. e. explaining) the general process of orientation?

The purpose of the following text and Tables is to distinguish, but not to judge, the varieties of orientations exemplified by tropisms and stratagems, scientific experiments and mystical revelations, artistic expressions and dogmatic interpretations of sacred texts, rigorous proofs of mathematical theorems and psychotic delusions and hallucinations.

Functional explanation and theory of action.
All the four modes of explanation[5] that meet the requirements of scientific epistemology have been employed in attempts to construct theories of behavior, but the greatest theoretical advance has occurred in the application of the functional mode of explanation to behavioral phenomena. The functional mode has been precisely formulated and successfully applied in physics. Cybernetics, decision

and game theory, information theory, and other important systems theories are functional theories.  Although there exists considerable confusion as to the exact format, functional formulations in biology, psychology, economics and sociology possess sufficient degree of similarity to make the work of different authors highly cumulative. For this and other reasons, the functional mode of explanation has been chosen as the format for the present theoretical work.

Theories employing the functional mode of explanation are usually called Theories of Action.  Since both the term "functional explanation" and the term "Theory of Action" evoke a variety of only partially overlapping associations, the meanings of these terms, as intended in this paper, are briefly indicated.

The functional mode explains the behavior of a system as Action, i. e. as the distribution of a definite amount of energy (quantum of action) through time and space subject to definite constraints. The action of the system is described as its transition from state P to state Q (within a definite time interval $t_i \rightarrow t_f$) along a particular trajectory (a locus of points in state-space, with time as parameter).  The fundamental question for the theory of action is which particular trajectory, among all the possible trajectories, actually describes the action of the system. The method is a search for variables that maintain extrema (constants, maxima, or minima) during the transition from $P(t_i)$ to $Q(t_f)$.  The aim of the theory of action is the specification of a set of extremum principles (and of intervening variables governed by them) that is sufficiently powerful to select a unique trajectory compatible with any given set of general boundary conditions (ranges of values of independent variables).

The general theory of action is envisaged as emerging through the integration of three special theories:  (1) an orientation theory, (2) a decision theory, and (3) a motivation theory.  Each special theory would be concerned with a different (special) fundamental problem in the explanation of action, employ different dependent variables, and invoke a different special extremum principle to explain a different part of the total variation in action.[6]

In an initial "kinetic" formulation of the general theory, the special theories would provide functional boundary conditions (ranges of admissible values of intervening variables) for each other. It is with a kinetic type of a general theory of action in mind that the special theory of orientation is being constructed. A "feedback" formulation of the general theory is anticipated as the intermediate step toward a "dynamic" general theory. A general theory, in any one of the three formulations, would also constitute a theory of development. [7]

A preliminary definition of orientation.

Although the idea of "orientation" seems intuitively simple and obvious, (and is invoked in most theories of behavior), it is not easy to give a definition of orientation acceptable for technical purposes. As is often the case, the central concept of a theory can be defined only by the theory itself.

In the absence of a substantive technical definition, the term "orientation" can be characterized by: (1) a summary of relevant usage, i.e. a dictionary definition; (2) a listing of the phenomena the term is intended to subsume, i.e. a denotative definition; and (3) a program for the use of the term in the theory, i.e. a quasi-syntactic definition.

(1)  The relevant usage is summarized in an authoritative technical dictionary as follows:

> "orientation: n. 1. the discovery or knowledge of where one is and where one is going, either literally in space and time, or figuratively in relation to a confusing situation or a puzzling problem, or to people and personal relations. The orientation is cognitive when it consists chiefly in knowing the situation; positively or negatively cathective when it consists primarily in feelings; evaluative when comparisons are made and the relation of the situation to personal goals is brought out. --Partial syn. insight. 2. the discovery of what or who one is; =self-orientation. --Syn. autopsychic orientation (rare), self-insight. 3. turning toward a source of stimulation (cp. tropism) or in a prescribed direction, literally or figuratively. 4. a set toward a certain stimulus, or a predisposition toward certain behavior patterns. 5. the direction taken by something. 6. a very general point of view, not necessarily verbalized, which helps to determine acceptance or rejection of scientific postulates, hypotheses, and

methodologies.  7. the process of helping a person to an orientation in any of the above senses: orientation program. --v. orient (not orientate). --adj. oriented (not orientated). " [8]

(2)  As presently envisaged, the theory of orientation would be a special theory, in that it would seek to explain only a portion of the total variation in behavior as a function of the orientation of an organism. [4] It would subsume most of the phenomena traditionally discussed as perception, cognition and thinking as instances of epistemic orientation.  It would also subsume many aspects of emotion (but not of motivation), and nearly all prudential, moral, ethical, normative and other forms of valuation (but not of decision) as instances of telic orientation.  In principle, it should explain not only the orientation of known organisms and collectivities, but also of specifiable alien and mechanical systems exhibiting behavior.

(3)  The description of the use of the term orientation in the theory can at this point be only programmatic.  The remainder of the paper is, in one respect, an attempt to provide an adequate syntactic definition.  The definition given below is quasi-syntactic, since it invokes two additional undefined terms, i. e. "experience" and "definition of action-world."  While a vast literature exists in philosophy, psychology and sociology on the meaning of those terms, technical definitions suitable in the present context have not been found.

Orientation is that operation on the manifold of experience by which it is transformed into a definition of an action-world.

The manifold of experience is the basic operand of orientation. It can be characterized as an amalgam of potential objects and undifferentiated properties.  In the context of the theory of orientation, the manifold of experience is regarded as an indeterminate field of a system, i. e. the objects, variables and relations of the system are unspecified.

The definition of an action-world is the final transform (or product, or outcome) of orientation.  It is a system made determinate by the specification of:  (1) the dimensions of the world; (2) the structure of the world; (3) the boundaries of a focal region, i. e. by a definition of the actor; and (4) the available action-lines, i. e. by a definition of the situation.

The above quasi-syntactic definition of the basic concepts "experience," "definition of an action-world, " and "orientation" provides the basic format for the analysis of the process of orientation. Of the three concepts, the first two have been characterized, however sparsely and informally. At this initial stage of analysis they act as the only defining terms for the as yet unanalyzed concept "orientation. " This state of the theory is represented in Figure 1. The definition of an action-world and the systems of discourse.

How many different types of orientation are there? This problem was investigated by a phenomenological analysis of the most directly accessible phase of the process of orientation--its outcome, i. e. the definition of an action-world. The results were reported in the first working paper[3] and are summarized and placed in context in this section.

As suggested in the preceding section the definition of an action-world can be treated as a system, which is made determinate to the extent that its constituents are specified. We shall call the act of specification an utterance (regardless whether it is public or private, conscious or unconscious, explicit or implicit, manifest or inferred).

A minimal list of necessary utterances that would constitute the definition of an action-world can be obtained by combining the irreducible facets of the concept "action-world" with the necessary components of the format "system. " Such minimal list is given in Table 1.

The column headings of Table 1 give the minimal components of the format "system, " the row headings the irreducible facets of "action-world. " The table yields explicitly the basic constituents needed for the definition of an action-world. These were already mentioned, more informally, in the previous section. The entries constitute the basic formal terms and concepts that will be employed and elaborated in subsequent text and tables. Row (T) entries ren-

der equivalent relevant set-theoretic, topological, and socio-psycho-
logical basic terms. [9]

In the presently relevant context, it is not necessary to distinguish between the kind of utterances needed to specify constituents 1a, 1b, and 2, respectively.   These and further such distinctions are made in subsequent tables.   Instead, we shall refine the
classification of utterances along the vertical axis of Table 1.
There, the set of all action-world defining utterances has been partitioned into two equivalence classes.   Class E of all epistemic utterances, and class T of all telic utterances.   Further phenomenological analysis of each class of utterances yields a partition of
class E into six equivalence sub-classes, and a partition of class T
into six equivalence sub-classes.   Altogether, then, twelve equivalence classes of action-world defining utterances have been distinguished.

An utterance is an accepted statement.   "Accepted" is a value
assigned to a statement by some operation of judgment or validation.
The operand of validation is a parent set, containing all available
statements.   Since utterances differ from available statements only
in having the value "acepted," we can induce from the partition of
utterances a corresponding partition of the set of available statements.   There are at least twelve classes of statements in the partition, each being a parent set of the corresponding class of utterances.   Each of the twelve classes of statements is a system of
statements, and has been labeled a system of discourse.   The description of the phenomenological characteristics of the statements in
the different systems of discourse was the topic of the previous
working paper, [3] and is summarized in the entries of Tables 5a - 8.
The fundamental phenomenological similarities and differences between the twelve systems of discourse are accounted for by a set
of three distinctions, which jointly yield a typology given in Table 2.

Table 2 displays the cross-partition of systems of discourse
into four equivalence classes (each of which contains three systems
of discourse), that is generated by the distinction between epistemic

and telic systems on the one hand, and between simple and combined systems on the other hand.

Telic systems differ from epistemic systems in the way indicated in Table 1.  Only in telic discourse a definition of an actor is given, and the meaning of all statements depends on (a) their relevance to that definition (situation or residual environment) and (b) their compatibility with that definition (eu-or dysfunctional).

Both simple and combined systems of discourse contain statements capable of specifying the variables, values and relations of a world.  Combined systems, however, contain two additional types of statements:  (1) correspondence statements (that combine the meaning of two or more terms from two different simple systems) and (2) transfer statements (that project the meaning of a statement from one simple system into another simple system).

Additional phenomenological criteria generate equivalence classes identified in Table 2 (and in subsequent tables) by the symbols 1, 2, 3, 2&1, 2&3, and 1&3.   Each of these classes contains two systems, one epistemic and one telic.   The phenomenological criteria of equivalence cannot be adequately summarized here, except by saying that statements in the two different systems of the same class are equivalent under a subjectifying transformation (epistemic to telic), or under its inverse, an objectifying transformation (telic to epistemic).   The symbols identifying the combined systems indicate which two simple systems each combines.

Three results of the phenomenological analysis are relevent to what follows:

(1)   All the kinds of utterances that are necessary and sufficient for the definition of any action-world can be accounted for by a twelve-fold typology yielded by a cross-partition based on three sets of phenomenological distinctions.

(2)   A one-one correspondence under a common transformation has been established between the six telic systems of discourse and the previously much better understood six epistemic systems of discourse.

(3)   The formulation of the process of orientation has been advanced from the state represented in Figure 1 to the state represented by Figure 2.

Figure 2 indicates the following:

(a)   The final transform of experience, the definition of an action-world, has been fully analyzed.   It consists of utterances, i. e. accepted statements.   The minimal requirements are indicated in Table 1.   In subsequent tables the distinctions along both the horizontal and the vertical axis are refined to an extent regarded as sufficient to account for all the variants of world-defining utterances.

(b)   An intermediate transform of experience, the systems of discourse, has been postulated.   It consists of available statements. It has been fully analyzed phenomenologically.   The resultant typology of statements is presented in the entries of Tables 2 - 6c.

(c)   The previously unanalyzed operator, "orientation," has been partitioned into two classes of operators.   Class $0_2$, tentatively called the systems of validation, acts on the systems of discourse as its operand and assigns the value "accepted" to some available statements, thus transforming them into utterances in the definition of an action-world.   Class $0_1$ operators presumably act on the initial operand "experience" and transform it into systems of discourse.

The systems of orientation.

Given the results of the phenomenological analysis of the systems of discourse, their formal characteristics will be considered in this section.

The results reported below were obtained in the following order:   (1) the formal characteristics of the systems of discourse were analyzed.   (a) A set of formal characteristics common to all systems of discourse was abstracted.   (b) Distinguishing formal characteristics of the various systems of discourse were identified up to a point of refinement that matched the articulation of the phenomenological typology.   (2) Assuming that the systems of discourse are a transform of "experience," the first phase of the process of orientation was reconstructed by postulating a set of systems of orientation as the Class $0_1$ operator in Figure 2. The systems of orientation consist of two kind of operators: (a) pandemic operators, which generate all the common formal properties of all systems of discourse, [10] and (b) endemic operators, which select the distinguishing formal characteristics of each of the twelve different systems of discourse.

For the sake of simplicity and brevity, the exposition of the results will be in the reverse order: First, we shall identify the various operators in the systems of orientation, and then indicate which formal and phenomenological characteristics of the various systems of discourse they generate.

Since all the twelve systems of discourse are alike in that they consist of statements, and since all statements are either (a) expressions of relationships between objects and properties, or (b) expressions of relationships between properties, it will be useful to keep in mind the following definitions:

Let an element be an object or a property. Let a statement be a relation on an ordered set of two elements. [11] The set of all the first elements that satisfy a relation is the domain of the relation. The set of all the second elements that satisfy a relation is the range of the relation. The set of elements formed by the union of the domain and the range of a relation is the field of the relation. The set of all elements in a system is the field of the system. The set of all relations in a system is the structure of the system.

Table 3 lists all the generative and selective operators that appear necessary to transform experience into all the kinds of statements that, according to the previous formulation, are available in the twelve systems of discourse.

The pandemic operators generate objects, properties and relations. Objects are generated by attaching an arbitrary label (tag) to any segment of experience by the denoting operator 1.1a (an index set). By repeating the same operation, objects may be arbitrarily subdivided or aggregated. This automatically segregates the unlabeled remainder of experience as the background, or the environment of the objects.   There is a number of well-known philosophical problems in the pandemic analysis of the manifold concerning the nature of objects, the possibility of exhausting experience by indexing, and the possible special status of general and specific boundary constants (such as time and space).   At this stage of the formulation these problems are bypassed and operators 1. 1b,  1. 2a and 1. 2b are included for future consideration.

The second type of pandemic operators differentiate properties and collect them by the process of abstraction into four kinds of sets--fields, variables, relations and systems.   It is assumed that there is an indefinite number of potential properties and relations, and the problem of classification is avoided by disregarding the substantive characteristics of the properties and relations generated by type 2 operators.

The endemic operators select among the properties and relations generated by type 2 operators on the basis of formal characteristics.

Type 3 operators restrict the meaning of statements.   Type 3. 1 operators restrict the meaning of statements to, respectively, the concrete, the abstract or the formal aspects of properties. Type 3. 2 operators distinguish between statements according to the source of their meaning.   Jointly, the combinations of type 3 operators that are given in Table 4a are sufficient to generate the three classes of simple systems of discourse.

Type 4 operators extend the meaning of statements to encompass aspects of properties, and domains and sources of relations of two simple systems at a time.   This is sufficient to generate the three classes of the combined systems of discourse, as shown in Table 4b.

Type 5 operators distinguish between the epistemic and the telic systems.   For the sake of exposition we shall consider the construction of the telic system as if it involved a subjectifying transformation of an already constructed epistemic system.   It should be noted, however, that the formulation of the operators provides as well the inverse objectifying transformation (telic to epistemic). It also allows for the direct generation of both telic and epistemic systems from experience. No genetic implications should be read into the present manner of presentation.

In Table 1 the minimal constituents of an epistemic and a telic world were juxtaposed and the effect of the subjectifying operation can be read as the transformation of Row (E) into Row (T). The minimal properties of an epistemic world constitute its dimensions, i. e. the coordinates of an (epistemic) descriptive space. Operator 5. 1 emphasizes one or several of the dimensions (or other variables) and thus selects a set of essential variables, i. e. the coordinates of a telic hyperplane (evaluative frame of reference). Telic systems contain only relations (statements) whose range consists exclusively of telic properties (i. e. values of essential variables, or points in the telic hyperplane, or telic values). The range of epistemic relations (statements), on the other hand, is not restricted beyond the restrictions by operators 3. 1 on aspects of the field common to both epistemic and telic systems of the same type.

Operator 5. 2 selects one or more values of an essential (range) variable as critical values. These polarize the essential variable by stating the bounds of a critical interval (containing eutelic values) and its complement (containing dystelic values). The intersection of the critical intervals on all the essential variables defines the focal region on the telic hyperplane. The focal region constitutes a definition of the actor. [12]

In the epistemic system, the state of the organism is represented by a point in description space. When projected on the evaluation hyperplane, does it, or does it not fall into the focal region, i. e. within the definition of the actor? The possibility of this comparison is a prerequisite of a conceptualization of motivation and decision. It also provides the scheme within which one can propose as the principle of action the imperative of authenticity: act as to minimize the descrepancy between your state as an organism and your definition as an actor!

The epistemic systems of orientation generate statements about the conditions of action.  The state of the organism at a given time and the structure of the epistemic world define jointly the environment of the organism and determine its admissible trajectories (in epistemic space-time).  A functional projection by operator 5.1 of the above on the telic hyperplane defines the actor's situation.  Eufunctional and dysfunctional action-lines can now be distinguished. In this fashion the telic systems generate statements that are rules of action.  This is illustrated by a telic world-graph in Figure 4.

The operators listed in Table 3, when appropriately combined into twelve systems of orientation, can generate the formal and phenomenological characteristics of the twelve systems of discourse, as shown in Tables 5a-6c.  This advances the formulation of the process of orientation to a state shown in Figure 3.

Systems of validation.

As shown in Figure 3, the overall features of the process of orientation have now been formulated as follows:  The initial operand, experience, is a (Jamesian) manifold conceptualized as an amalgam of potential objects and undifferentiated properties and formalized as an indeterminate field of an unspecified system.  The operators of the systems of orientation analyze experience into elements (objects and properties) and order the elements into collections (aggregates, sets, fields, variables, relations, systems).  The results of this transformation are available statements organized in twelve systems of discourse.  Next, systems of validation operate on the systems of discourse and accept some of the available statements, thereby transforming them into utterances.  A system of utterances constitutes the definition of an action-world.  This yields the final transform of experience, a specified and relatively determinate system.

Each system of discourse contains statements, which, if accepted, yield a different kind of knowledge of the action-world.  The phenomenological analysis distinguished twelve types of knowledge,

each made available by one system of discourse. The formal anal-
sis of the systems of orientation explains how the differences in the
types of knowledge are generated by the different combinations of
endemic operators. This is shown in Table 7, which gives the phenomenological characteristics of the twelve types of knowledge, and
in Tables 4a and 4b, which indicate the formal constraints on the
meaning of the different types of knowledge.

Table 7 gives us a point of departure for the analysis of the
last unanalyzed mechanism of the process of orientation, the systems of validation. Twelve different dimensions of validity are postulated, each associated with one of the systems of discourse. Each
system of validation will then need to contain operations that are
necessary and sufficient to determine the acceptability of a statement with reference to the relevant dimension of validity.

The four major classes of statements are again distinguished
in terms of the criteria of their validity. Knowledge provided by
simple systems claims a single sphere of validity, while knowledge
provided by combined systems claims a double sphere of validity.
The validity of epistemic statements is absolute, i.e. depends on
the judgment that they are independent of any definition of an actor.
The validity of telic statements is relative or contingent, i.e. depends on the judgment that they are relevant to, and compatible or
incompatible with, a specific definition of a given actor.

The Principle of orientation.

To advance beyond the conceptualization of orientation to a theory, an explanatory principle governing the process of orientation
and determining the relative contribution of its phases and types
must be postulated.

At the present time, some principle of management of uncertainty is being considered. While many problems remain to be
solved, [13] it appears plausible that it could be formulated as a minimal principle, i.e. a principle of reduction of uncertainty. Accord-
ingly, from all the available statements such definition of an action-
world will be accepted as has minimum possible uncertainty.

The principle of reduction of uncertainty leads to the postula-
tion of gradients of uncertainty, associated with each system. The
gradients of uncertainty of the telic systems are shown in Table 8.

Telic gradients of uncertainty are polarized by operator 5.2. Uncertainty is at a relative maximum when it is not possible to determine whether a given state of the organism is within or outside the focal region (its definition as the actor).   Evidence of compliance (+) or violation (-) of the definition of the actor reduces uncertainty equally toward a relative minimum.   This formulation also relates the theory of orientation to the theories of motivation and decision.

In the dynamic formulation of the theory of orientation the gradients of uncertainty will be arranged into uncertainty levels, as follows:

Level 1:   all aggregates of objects
Level 2:   all substantive properties and relations
Level 3:   all simple systems
Level 4:   combined systems 2&1 and 2&3
Level 5:   combined systems 1&3

The higher the level, the steeper the gradient of uncertainty, and therefore the greater the potential for uncertainty generation and reduction.

Some possible applications.

Concurrently with the formulation of the theory of orientation, attempts are being made to test it.   Five possible tests of the adequacy of the formulation are now being explored:

(1)   Its internal consistency, by an attempted axiomatization.

(2)   Its theoretical compatibility with (a) the attempted formulation of a general theory of action, and (b) with the attempted reformulations of motivation and decision theories as special theories of action.

(3)   Its philosophical consequences for the hitherto normative doctrines of epistemology and ethics.   Would it be possible to outline a behavioral epistemology and teleology that would not be normative?

(4)   Its empirical power, in three areas of substantive interest to the author and his collaborators: (a) One is in the social psychology of individual development.   (b) Another is in the area of psychopathology, where an attempt is made to use the theory to explain the major functional psychoses as "disorders of orientation."   (c) The last area is the study of ideology, especially its role in the rise

and development of the various types of collective behavior.

(5) Its algorithmic adequacy as a guide to the programming of higher epistemic and telic processes on computers. This is a test by "simulation," in that a demonstration of its ability to generate artificial intelligence and artificial conscience in automata could be regarded as enhancing the plausibility of the theory.

## Notes

1.   This paper is a substantially revised and updated version of a research report which was undertaken, in part, at the IBM Th. J. Watson Research Center during the summer of 1963 with the support of contract AF19(628)-2752. The author was then affiliated with Rutgers University.

2.   The present version incorporates results obtained in the fall of 1964 at Cornell University in the research context of the Pro Program in Social Systems Analysis.

3.   The original version of the first working paper in the theory of orientation was read at the annual meeting of the American Association for the Advancement of Science, Philadelphia, Pa., December 30, 1962, under the title: "Formal Analysis of Ideological Components of Behavior." A revised and expanded version will appear under the title: "Systems of Discourse" in a book edited by Anthony Leeds and planned for publication in 1966.

4.   Since no generic term is available to characterize all behaving systems (organisms, machines, and collectivities formed of combinations of organisms and/or machines) the term "organism" will be used hereafter as a generic term for all such systems.

5.   Scientific explanations can be classified into four relatively well understood types. These are the deterministic, the functional, the genetic-comparative, and the stochastic mode of explanation.

6.   For a preliminary formulation of an approach to a special theory of motivation, cf. Jung, R., "Introduction to Part II" in Klausner, S. Z. (Ed.), The Quest for Self-Control, New York: The Free Press, 1965, pp. 121-125. A Working Paper outlining a com-

patible special theory of decision is being prepared for publication.

7. A general approach to the construction of a theory of development has been discussed in Jung, R., Analysis of Psychosocial Development, Unpublished Ph. D. dissertation, Harvard, 1962.

8. Quoted from English, H. B., and English, Ava C., A Comprehensive Dictionary of Psychological and Psychoanalytic Terms, New York: Longman, Green, 1958, pp. 363-364.

9. The terminological melange in Table 1 ff. results from a search for concepts compatible with a unified formulation. The basic terms in Table 1, row (T), draw on relevant functional formulations of physics, of cybernetics, and of sociology. The most directly relevant sources are:

Yourgrau, H., and Mandelstam, S., Variational Principles in Dynamics and Quantum Theory, London: Pitman, 1955, Chapters II & III.
Asby, W. R., "The Set Theory of Mechanism and Homeostasis," in Bertalanffy, L, and Rapoport, A. (Eds.), General Systems, Vol. IX, Bedford, Mass.: Society for General Systems Research, 1964, pp. 83-97, esp. 91-92.
Parsons, T., Shils, E. A., with Olds, J., "Values, Motives, and Systems of Action," in Parsons, T., and Shils, E. A. (Eds.) Toward A General Theory of Action, Cambridge, Mass.: Harvard University Press, 1954, esp. pp. 53-68.

10. Pandemic operators appear sufficient to generate experience.

11. The relations may be on n elements, but the discussion of binary relations is simpler.

12. Each telic system produces different definitions of the actor. In a complete definition of an action-world, at least one definition of each type is incorporated. See pp. 80-81.

13. Among the problems that must be solved are: (1) Are there two types of uncertainty, semantic and syntactic, corresponding to the signification meaning of <property, object> relations and the structural meaning of <property, property> relations? (2) What is the source of uncertainty? If it is to be attributed to experience, then experience itself must be reformulated as the product of the pandemic operators.

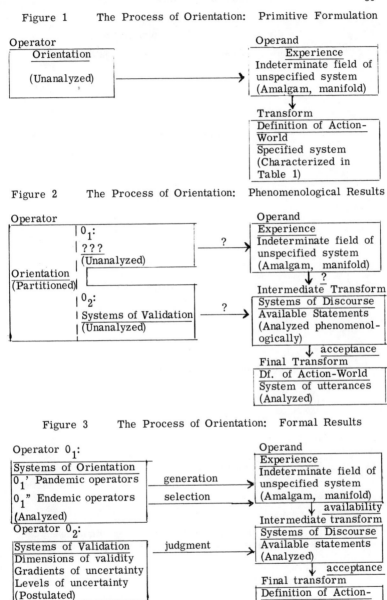

Figure 1    The Process of Orientation:    Primitive Formulation

Figure 2    The Process of Orientation:    Phenomenological Results

Figure 3    The Process of Orientation:    Formal Results

Figure 4.　　A Telic World-Graph

(of the situation of an organism in state $S_O$)

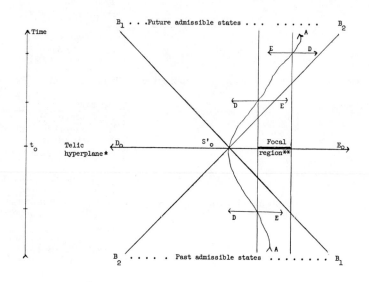

| $t_O$ | = Now |
|---|---|
| $S'_O$ | = State of the organism at $t_O$, in a functional projection on the telic hyperplane, i.e. the present telic value of the state of the organism |
| A | = An admissible action-line |
| $B_1, B_2$ | = Projections of epistemic boundaries on admissible states, given a structure of the epistemic world and a state of the organism $S_O$ |
| D | = Dysfunctional direction of action at a given point in space-time |
| E | = Eufunctional direction of action at a given point in space-time |
| * | N-dimensional hyperplane represented by one coordinate |
| ** | The counterintuitive simplifying assumption is made that the definition of actor (focal region) does not change over time |

Table 1    The Definition of an Action-World:    Minimal Constituents

| System | (1a) Variables | (1b) Their Value | (2) Relations on the Variables |
|---|---|---|---|
| **Action-World** | | | |
| Epistemic World (E) | Dimensions  Epistemic space | State  Epistemic event | Structure  Epistemic trajectories |
| Telic World (T) | Essential variables  Telic hyperplane  (Evaluative frame of reference) | Critical intervals  Focal region  (Definition of actor) | Functional projections  Action lines  (Definition of situation) |

Table 2    The Twelve Systems of Discourse:    Phenomenological Typology

| Type of System | | Epistemic | Telic |
|---|---|---|---|
| Simple | 1 | Accounting | Aesthetic |
| | 2 | Conceptual | Moral |
| | 3 | Explanatory | Religious |
| Combined | 2 & 1 | Methodological | Pragmatic |
| | 2 & 3 | Theoretical | Ethical |
| | 1 & 3 | Representational | Expressive |

Table 3          Transforms of Experience Under Generative and
                          Selective Operators

| Operators | Transforms |
|---|---|
| $0_1'$: Pandemic operators | |
| 1.    Object generators | Objects, aggregates of objects<br>(strings, lists) |
| 1.1  Denoting operators | Inventory |
| | Value;   aggre-  type,  universal<br>             gate,          interval |
| 1.1a Index (tag, label) | Individual; sample, class, population |
| 1.1b General boundary<br>     (complement) | Condition; context, situ-  environ-<br>                    ation,  ment |
| 1.2  Specific boundary<br>     operators | Chronography and chorography |
| | Value;   interval, type,   universal<br>                          interval |
| 1.2a Time | Date;   period,  stage,  epoch |
| 1.2b Space | Location; area,   region, universe |
| 2.    Set abstractors | Substantive characteristics |
| 2.1  Common property | Variables (V) |
| 2.2  Common order | Relations (R) |
| 2.3  (2.1 & 2.2) | Fields, spaces, systems |
| $0_1''$: Endemic operators | Formal characteristics |
| 3.    Meaning restrictors | |
| 3.1  Aspect restrictors | |
| 3.11 Concrete ⎫ | Concrete ⎫ |
| 3.12 Abstract  ⎬ values, fields,<br>3.13 Formal    ⎭ variables | Abstract  ⎬ properties<br>Formal    ⎭ |
| 3.2  Domain quantifiers | |
| 3.21 (∃) quantifier | Particular ⎫ statements |
| 3.22 (∀) quantifier | Universal  ⎭ |
| 3.3  Source restrictors | |
| 3.31 Endosystemic (syntactic)<br>     implication | Analytic ⎫<br>         ⎬ statements |
| 3.32 Any other source | Synthetic ⎭ |
| 4.    Meaning extendors | Combination statements |
| 4.1  Correspondence rules | Correspondence statements |
| 4.2  Transfer procedures | Transfer statements |
| 5.    Subjectifying operators | |
| 5.1  Emphasizing operator | Essential variables, telic hyper-<br>plane, telic values; functional vs.<br>non-functional relations (statements) |
| 5.2  Polarizing operator | Critical values and intervals, focal<br>region, eutelic vs. dystelic values;<br>eufunctional vs. dysfunctional re-<br>lations (statements) |

Table 4a    Simple Systems of Orientation

| System Number | Variables, Values and Fields | Description or Evaluation Space Relations | System Relations |
|---|---|---|---|
| 1 | Concrete | ( A∃ ) Analytic & Existential | ( S∃ ) Synthetic & Existential |
| 2 | Abstract | ( A∃ ) Analytic & Existential | ( SV ) Synthetic & Universal |
| 3 | Formal | ( A∃ ) Analytic & Existential | ( AV ) Analytic & Universal |

Table 4b    Combined Systems of Orientation

| System Number | Variables, Values and Fields | Description or Evaluation Space Relations* | System Relations** |
|---|---|---|---|
| 2 & 1 | Abstract & Concrete | $( A∃ )_2 \equiv ( A∃ )_1$ | $( SV )_2 \Longleftrightarrow ( S∃ )_1$ |
| 1 & 3 | Abstract & Formal | $( A∃ )_2 \equiv ( A∃ )_3$ | $( SV )_2 \Longleftrightarrow (AV )_3$ |
| 1 & 3 | Concrete & Formal | $( A∃ )_1 \equiv ( A∃ )_3$ | $(S∃)_1 \Longleftrightarrow ( AV )_3$ |

\*   " ≡ " means rendered equivalent by a correspondence rule.
\*\*  " ⟺ " means mutually derivable by appropriate transfer procedures.

Table 5a    Epistemic Systems of Discourse: Simple Systems

| System Number | Name of System | Variables | Values of Variables | Relations |
|---|---|---|---|---|
| 1 | Accounting | Observers | States of observers | Patterns |
| 2 | Conceptual | Class concepts | Element concepts | Propositions |
| 3 | Explanatory | Variable symbols | Value Symbols | Functions |

Table 5b    Epistemic Systems of Discourse: Combination Statements*

| System Number (X & Y) | Name of System | Correspondence Rule | | Transfer Procedure | |
|---|---|---|---|---|---|
| | | $(V \epsilon\ X) <= (V \epsilon\ Y)$ | $(V \epsilon\ X) => (V \epsilon\ Y)$ | $(R \epsilon\ X) <= (R \epsilon\ Y)$ | $(R \epsilon\ X) => (R \epsilon\ Y)$ |
| 2 & 1 | Methodological | Interpretation as indicator | Operational definition | Inductive generalization | Test of hypothesis |
| 2 & 3 | Theoretical | Interpretation | Lexical definition | Deductive inference | Explanation |
| 1 & 3 | Representational | Representation | Symbolization | Production | Simulation |

Table 5c    Epistemic Systems of Discourse: Combined Systems

| System Number (X & Y) | Name of System | Variables** | Values of Variables | Relations |
|---|---|---|---|---|
| 2 & 1 | Methodological | $(V_i \epsilon\ 2) \equiv (V_j \epsilon\ 1)$ | Facts | Hypotheses |
| 2 & 3 | Theoretical | $(V_i \epsilon\ 2) \equiv (V_j \epsilon\ 3)$ | Constructs | Laws |
| 1 & 3 | Representational | $(V_i \epsilon\ 1) \equiv (V_j \epsilon\ 3)$ | Artifacts | Models |

*   X, Y = respective simple systems; V = variables and implicitly their values v; R = relations, arrow indicates the direction of mapping.

**  No generic concepts for the three types of statements of correspondence between variables in two different systems are available in current terminology.

Table 6a　　Telic Systems of Discourse: Simple Systems

| System Number | Name of System | Subjectifying Operation | Critical Values | Focal Region (df. of Actor) | Functional Relations |
|---|---|---|---|---|---|
| 1 | Aesthetic | Separation | Thresholds | Body-Self | Emotions |
| 2 | Moral | Commitment | Standards | Role | Rules of conduct |
| 3 | Religious | Individuation | Values | Condition | Rules of importance |

Table 6b　　Telic Systems of Discourse: Combination Statements*

| System Number (X & Y) | Name of System | Correspondence Rule | | Transfer Procedure | |
|---|---|---|---|---|---|
| | | $(V \in X) <= (V \in Y)$ | $(V \in X) \equiv >(V \in Y)$ | $(R \in X) <= (R \in Y)$ | $(R \in X) = >(R \in Y)$ |
| 2 & 1 | Pragmatic | Interpretation as sanction | Cathexis | Prudential generalization | Pragmatic experiment |
| 2 & 3 | Ethical | Interpretation as duty (responsibility) | Identification (loyalty) | Categorical command | Categorical judgment |
| 1 & 3 | Expressive | Representation | Symbolization | Incarnation | Sublimation |

Table 6c　　Telic Systems of Discourse: Combined Systems

| System Number | Name of System | Subjectifying Operator | Critical Values | Focal Region (df. of Actor) | Functional Relations |
|---|---|---|---|---|---|
| 2 & 1 | Pragmatic | Engagement | Constraints | Regulator | Prudential rules |
| 2 & 3 | Ethical | Calling (Appointment) | Norms | Office | Categorical rules |
| 1 & 3 | Expressive | Posture | Stigmata | Status | Rituals |

* X, Y = respective simple systems; V=variables and their values, implicitly also telic ranges, critical values and focal regions; R = relations; arrow indicates the direction of mapping.

Table 7    Dimensions of Validity

| Sphere of Validity | (System) | Absolute with respect to what is: (Epistemic) | Contingent on (relative to) the actor's: (Telic) |
|---|---|---|---|
| Single | (1) | actual | existence |
|  | (2) | possible | identity |
|  | (3) | necessary | destiny |
| Double | (2&1) | real | task |
|  | (2&3) | ideal | mission |
|  | (1&3) | permissible | image |

Table 8 Telic Systems: Gradients of Uncertainty

| System Number | Name of System | Uncertainty | | Polarized* | |
|---|---|---|---|---|---|
| | | MAX | MIN | MIN (+) | MIN (−) |
| 1 | Aesthetic | Irritation | Emotional clarity | Pleasure (Well-being) | Pain (Misery) |
| 2 | Moral | Functional insignificance | Functional significance | Eufunctional significance | Dysfunctional significance |
| 3 | Religious | Absurdity | Meaning | Comprehension | Mystery |
| 2 & 1 | Pragmatic | Impotence | Competence | Mastery | Endurance |
| 2 & 3 | Ethical | Anomie | Normative determination | Prescription | Proscription |
| 1 & 3 | Expressive | Isolation (Loneliness) | Integration | Solidarity | Antagonism |

* Polarization of the uncertainty gradient is one of the results of the partitioning operation; cf. Table 3. (5.2)

I am indebted to Professor Burkart Holzner for a suggestion that lead to the retention of an original insight concerning a polarity of "Integration" and to its generalization to all the six systems.

## II B.   A Model for the Process of Learning to Comprehend

M. Kochen and L. Uhr
University of Michigan

### A.   Introduction

In a previous paper[1] we stressed the need for general learning capability (discovery and induction) to enable an entity with such capability to make leaps towards understanding items in a completely unlimited domain of discourse.   We started to outline the simplest kinds of automata with this capability, with a view toward implementing their behavior by computer programs (coded in IPL-V, COMIT and SNOBOL).   In this paper, we recapitulate, and develop the model somewhat further.

The entities (automata) in which we are interested are to process information inputs from their environments and take actions so as to continually improve their internal representations of these environments.   Prior to discussing the minimal capabilities that have to be built into such automata, we consider the way in which their internal representations (models of their relevant environments) are to be treated.

### B.   Internal Representations

We consider a universe characterized by a set of states, and a mapping of this set into itself.   Embedded in this (mechanistic) universe are: the automaton M under study and its relevant environment, E, also treated as an automaton.   All the input variables of M occur among the output variables of E, and all the output variables of M occur among the input variables of E.

Viewed abstractly as an automaton, M is characterized by a

set of internal state-variables, a set of input variables and a set of output variables; the values of all the output variables and the internal state variables at any time are co-determined by the values of all the input and state-variables at the preceding time. This assumes discrete time units. The co-determination is specified by a function called the behavior-function, $b_M$. This function is to evolve through continued interaction between M and E, according to a "value maximization" principle.

All possible values of M's internal-state variables constitute its set of states, or its state-space $S_M$. It is assumed that, for each point $s \in S_M$, there is a "value function" $v(s)$ which indicates whether s is a need-state or a highly valued state. The principle of "value maximization" is that $b_M$ will be such that the output-variables of M (responses, actions) lead to future states for which $v(s)$ is maximum. This is akin to planning. To make it possible, information about E, in terms of the values of M's input variables, must continually be sampled for better approximations to $b_E$, the behavior function of E.

This, however, is at a very abstract level. More concretely, inputs from E come both as direct perceptual stimuli and as linguistic stimuli. These have to be correlated and assimilated into a partial model, or internal representation, of the behavior of E. What are the logical constituents of such a model?

1.  Individual constants representing simple nominal concepts: $A_1$, $A_2$, . . . , forming a set A, which grows with time.
    Example: $A_1$ = Empire State Building, $A_2$ = Chrysler Building.
2.  Individual constants representing simple predicate concepts. These are of two types:
    (a)  Quality concepts: $F_1$, $F_2$, . . .
         Examples: $F_1$ = is taller than, $F_2$ = loves, $F_3$ = is green.
    (b)  Activity concepts: $G_1$, $G_2$, . . .
         Examples: $G_1$ = builds, $G_2$ = moves . . . from . . .to . . .
3.  X is a substantive concept if it is compounded from simple concepts using a set of substantive concept-operators applied

according to a set of rules.  These also vary with time and
are acquired through M–E interaction.

Example: Empire State Building and Chrysler Building = sky-
scraper?

4.    X is a predicate concept if it is compounded from simple concepts using a set of predicate concept-operators and rules for
applying them.

Example: (is green and is round) or (is square and is tall) or
(if not green, then tall and not square).

5.    X is an internally represented event or state of E if it is compounded of substantive and predicate concepts according to a
set of rules acquired through M–E interaction.

Example: $A_1 \ F_1 \ A_2$ = Empire State Building is taller than
Chrysler Building.   $A_2 \ G_1 \ A_1$ = Chrysler Building builds Empire State Building.

The rules of formation are continually reformulated so that
they continually improve in producing representations which are
meaningful and true.  If this is not the case, action based on inadequate representations will result in low-valued internal states.

The representations $A_1$, $G_1$, $A_1 \ F_1 \ A_2$, etc., are, of course,
points in the state space $S_M$.  For each such point, there will,
eventually, be a linguistic representation: $a_1$, $g_1$, $a_1 \ f_2$, etc.  Here
$a_1$ denotes the phrase for the Empire State Building, or one of its
near-synonyms; and, being a phrase, its homographs play a role also.  It is, perhaps, the homographs, metaphors, and other ambiguities of "natural" language which allow for its great flexibility and
fertility as a form for the representation of concepts.  (Incidentally,
it is probably these features of a "natural," living language which
furnish the greatest obstacles to economic automation).  Each of the
rules for compounding simple concepts into concepts, substantive
and predicate concepts  into event and state representations has its
counterpart in linguistics.

C.   Organization of Automata

We now turn to the question of how M converts raw sense data into information to update its internal representation of $b_E$. The fundamental distinction between M and E is that, in a sense to be made precise, E is immensely more complex than M. State transitions in E cause significant disturbances for M, but not vice versa.

It is useful to view all the output variables of E and M as n-valued variables. The n-bit number denoting the values of all n output variables of M at any time is called the output. The time sequence of such n-bit numbers at successive (discrete) times is called the output sequence of M.

To say that E is more complex than M is to say that E can generate output sequences that M cannot analyze (i.e. reconstruct the rules by which the output sequence of E was generated, not by imitation); also, that E can generate output sequences of complexity greater than that of any output sequence M can generate. Complexity of an output sequence is a function of the period (since, if M, E are finite-state automata, all their output sequences are periodic) and the symmetries, the regularities in the pattern of one period.

We view the automaton M to be organized as follows:

1. There is an input (output) register, holding up to a certain number of bytes, where a byte is an n-bit number denoting the input (output) at one time.

2. A storage/recall organ composed of variable length-cells with links to other such cells.

3. Attention registers at several levels. Those at level 1 hold contents of cells transferred directly from memory cells. The next level registers hold the contents transferred from cells at level 1.

D. Processes

The following elementary operations governing information flow among these registers seem to be needed, as a minimum.

1. Shifting the contents of the input (and output) register, destroying the oldest byte stored and allowing storage of the most recently arrived input.

2. Testing two byte-sequences for equality (Matching).

3.    Storing an item (a byte-sequence, or code for it) into a speci-
      fied cell or into the output register.

4.    Composing two items to form a third, as in composing two
      "words" to form a new "word" (e. g. tabletop).

5.    Associating two items relative to a specified type of relation
      (e. g. "red-light," "stop" relative to a prescribed action).

6.    Including an item in a class (Naming).

7.    Emitting output sequences.

The structure of the memory at any time can be pictured as a
graph in which the nodes are named things, compounds of things, or
classes, and the edges are relations (e. g. inclusion).[2] An M is
able to grow its store in the sense that it can add nodes and edges
to this graph as it needs to, according to a set of learning rules.

An M accepts inputs that enter into its input register from E.
It "recognizes" patterns in these inputs by shifting and matching
(testing for equality).   This is to check for patterns that M has previously encountered and found to be correlated with the attainment of
a high-valued state.

This process of singling out certain recurring patterns is equivalent to classifying such patterns and naming (or coding) the classe:
The resulting code points are new nodes of the memory graph.   The
pattern-forming process involves both the operations of composition
and naming (class inclusion) in addition to the shifting, matching and
storing operations mentioned above.

Links among the nodes in the memory graph are inferred from
temporal-spatial contiguities in the input sequence, using the association and storage operator.   Inferred links can also result from gen
eralizations in hypotheses formed by examining attributes common to
a number of nodes and applying the universal quantifier.

An M produces output sequences in order to effect state
changes in E which, in turn, will put M into highly valued internal
M-states at a future time.   Prior to emitting an output sequence,
however, M emits "proprioceptive" signals which determine   the
output.   Such signals denote readiness for action.   For example,

such a proprioceptive signal is associated with a link between the
node representing M itself in the memory graph and nodes representing objects on which M may act (e. g. apples, pears).

Values, v(s), may be attached not only to nodes of the memory graph but to node-pairs linked by such a proprioceptive signal or
to longer trails in the graph.    There is in M a mechanism for continually exploring such trails.    There are few radical jumps from
any one trail currently under examination to trails which differ
greatly (are far away).    If a trail leading to highly valued states is
discovered, the search is concentrated in the neighborhood of such
a trail.

E.   Simulation of Some   Processes of this General Sort

Algorithms for some of the processes mentioned on the last
two pages have been sketched, in some cases programmed, for the
most part formulated, and are being revised.    The first process involved in transforming new sense data--i. e. an input sequence--into
information to update M's internal map of E, its approximation to
$b_E$, can be illustrated as follows.

Consider a one-dimensional time sequence of bits (or letters
denoting bytes) as input to M.    The program simulating $b_M$ searches this input sequence in order to find patterns it has previously
learned and stored in its memory.    Each pattern that is found has
a set of information about it: the names of other, larger patterns of
which it is a part, the classes to which it belongs, and the names
of other patterns that it "implies."    These patterns correspond to
the nodes of the memory graph, and the dyadic relations of membership, "implication," etc., the edges.

The program combines patterns into more inclusive patterns
whenever possible.    For example, if the input sequence is a string
of letters forming an ordinary English sentence, the patterns found
first would be the individual letters.    Patterns composed of combinations of letters--probably words--would be found next.    Patterns
composed of words--probably idioms--might be found at the third
pass.    The process of finding letters would still continue, however,

at the second pass, and even through the third and fourth, etc.

At a certain pass, the program will have found a set of patterns (nodes on the graph), mixed together with a set of "unknowns." The latter are portions of the input sequence not yet matched by any of the patterns learned and stored so far.  The patterns (nodes) to which recognized patterns are linked are examined to check whether a proposed new linkage would be inconsistent with linkages already stored.   That is, ambiguities in a previously stored linkage would thus be detected and resolved by enlarging the patterns (composing them into larger patterns) involved in a revised linkage.   The patterns to which a given  already recognized pattern is linked are also examined to see whether class-patterns can be recognized among the classes to which the given pattern belongs.   Class-patterns are the generic names of classes, as a link from "poodle" to "dog. "

Such class-patterns, when found, indicate that the program would restructure the members of the several classes involved, according to a restructuring rule associated with the class-word found. These restructuring rules serve (a) to choose among alternate, hence ambiguous, transforms, (b) to break transforms into parts, and (c) to change the ordering of the transforms.   The program will often make several iterations through the routine that searches for class-words, because when a class-word is found that leads to the choice of a particular implication from among a set of alternates this implication, only now chosen by the program, will itself become a member of classes, some of which may help to form new class-words.

This general process of choosing a set of paths through the graph continues over several layers of depth, but in the same way outlined above.   In general, one can think of the context for a given pattern interacting with that pattern to determine the particular choices among its possible transforms.   The choices described above are made as a function of other patterns in the immediate neighborhood.   The overall flow is determined by more distant patterns, stored in what might be thought of as a "current memory, " or

"expectation" register. For example, the overall flow might be
from inputs in a particular language, say English, to written outputs
in a second language, say French. If the program has also learned
something about translating to German, and something about outputting in spoken French (e. g. by means of the phonetic alphabet), then
it must make the choices, through this more distant context, first
between the German and the French implicants (i. e. nodes to which
a given pattern is linked) of an English word-pattern found and then
between the written and the spoken representation.

Now the program outputs its response. Thus each response
can be a function of a complex set of subpatterns in the input sequence, and the judgments as to what are subpatterns (that is, as
to how to segment the input sequence) are made by the program as
a function of its learned memory.

If the program is given feedback, it attempts to learn something from this experience. (If it is not given feedback, it simply
continues to process the input string according to the set of inferences embodied in the present state of its memory graph. ) First
it looks for matches between the feedback sequence and the sequence
it has just output. If both a pattern and the pattern's relative position in the sequence match the program increases weights of a set
of inferences (that is, of node-edge-node trails of the graph) that
led to its outputting of that pattern. If the relative position is
wrong, the program makes a tentative inference, following a "learning rule" for permutations that would have corrected this mistake.
This new inference may involve a whole set of new constructions on
the graph--for example, several of the patterns may be made members of classes, possibly these classes will themselves be created,
and possibly a new class-word and a new restructuring rule about
the class-word will be created.

Next the program looks to see whether two different parts of
the feedback will combine into one of the pattern's outputs (that is,
whether that pattern has been chopped up into a discontinuous pattern). If such a situation exists, the program will follow "learning

rules" for discontinuous discourse and permutations to construct new
parts of the memory graph that would have corrected this mistake.

Next the program checks whether any of the patterns it output
are still not accounted for in the feedback. If, between two correctly
transformed patterns, there is one, or a set, of such patterns, the
program combines these up into a new whole, makes a tentative inference as to which of the other patterns in the string might be the
context that has led to the need for this new whole, and follows
"learning" rules for ambiguous situations and idiom formation in
order to construct new parts of its memory graph.

Finally, the program will be left with unknown portions of the
input string and unmatched portions of the feedback string. It will
make inferences as to how these are connected, and put into its memory graph a new 5-tuple for each item--the input pattern, the link
to other patterns, type of transform, initial weight of the inference,
and the output pattern.

At each of the above steps, when the program has made a
wrong inference, it weights down those parts of the graph that were
involved. When a part is weighted down below a certain cutoff, it
is erased from the graph and some statement is recorded in memory
so that this mistaken inference will not be repeated in the future.
When the program has made a correct inference, it weights up those
parts of the graph that were involved. Only when a part of the
graph has been weighted up beyond some point of "certainty" will
the program be free to add onto it by invoking one of the learning
rules alluded to above. Thus an inference will have to be confirmed
several times before further inferences can be based upon it.

Programs to implement certain parts of the above have already
been written by L. Uhr; some of the features described in the above
are also incorporated in Kochen's hypothesis-selection programs and
Uhr's pattern recognition and language learning programs. Design
of algorithms to implement some of the remainder are underway
jointly by Uhr and Kochen.

F.   Conclusions

Understanding is assumed to arise from the necessity of a
sufficiently complex and sensitive entity to assure its continued existence in an environment of immensely greater complexity than itself.   Though the term "understanding" is commonly used to denote
many functions and states, its use to denote the capability of making leaps toward a better understanding of anything with which an
unspecified domain of discourse confronts it seems to us the most
significant.

To achieve "understanders" and "intelligent" entities, we cannot expect to analyze domains and then design appropriate machines.
Rather we must try to get a grasp on methods for learning, which
we take to mean discovery and induction, and design machines that are
general learners.   These machines must be effectively coupled to sufficiently interesting environments.   By "sufficiently interesting" we
mean that the environment must be capable of transmitting the information we want the machine to absorb.   Environments will typically contain other machines, in which case language might be expected to
arise as a vehicle for sharing between machines whatever knowledge
each individual has amassed, and for arriving at a community of
interest and action.

Some simple aspects of the design of such a "cognitive" agent
appear to emerge.   To be connected with its environment, it must
sense and act.   To have any reason for absorbing information,
for becoming "cognitive" in the first place, the assumption of
a need-value system with a value-maximization principle or
something equivalent seems plausible.   From this, it must organize functions that can test to see whether desirable patterns, and indicants of desirable patterns, are there, and then it
must act appropriately--e. g. by placing new pattern-configurations
in its memory or onto its output tape.   Thus it must have as basics
such abilities as "match," "shift," "copy," and such connectors
as "compose," "associate," "include."

We hope that now it might be sensible to ask:   What is the
minimum complexity for such an entity that will allow it to try to

become an understander of a universe complex enough to contain it?

## Notes

1.    Appendix D to a RAND-report on "Computers and Comprehension," by Kochen, M., Mackay, D. C., Maron, M. E., Scriven, M., Uhr, L. entitled "On the Possibility of an Entity Which Can be Said to Understand."

2.    Graphs are best suited for depicting two-place relations. It is a fundamental problem to decide whether one should try to express all relations in terms of two-place relations, or extend the concept of a graph to allow representation of n-place relations.

## II C.  Methodology for Research in Concept-Learning

Robert Strom[1]

Harvard University

### Part I

I wish to introduce a change into the manner in which studies
of artificial intelligence in linguistic processing machines have been
carried out in the past.  Current and past methods of handling language learning and computer intelligence in general have involved
the system in pre-programming all the specific rules of the system
in advance.  If there is any variation at all during the course of the
operation of the program, such a change occurs in some numerical
parameter which exists in a prefabricated structure.  Thus, in the
Samuel program for checkers, the coefficients in the various polynomial strategies are allowed to vary during the program; in Simon's
GPS (General Problem Solver), it is the "effectiveness" rating of
each of the pre-formed rules toward eliminating differences between
a pre-set goal, and a current situation which varies.  In most language translation programs, no variation in strategy during the
course of the program has been established, and any that has been
planned (such as contextual analysis to eliminate the indecision between two translations when the text obviously points to one of them),
is in the form of rules designed by linguists to change some "probability" weight of choosing one of the ambiguous translations.  De-
spite the vast amount of time and effort spent on linguistic analysis
of individual languages  in order to decide beforehand exactly how
the computer program is to behave, the achievement level of the
"brute force" translation methods has been quite low.  Translations

105

are still incomprehensible and awkward.  If it becomes desirable to
try to translate a different language, say, Japanese, instead of Russian, a completely different group of linguists have to produce a
completely different linguistic analysis and formulate a new set of
complicated rules, which rules result in awkward translations.   The
further removed a language is from the Indo-European language category, the stranger the translations begin to sound.   By the time one
gets to some of the polysynthetic world languages, translations will
look ludicrous indeed, if done by present methods.   Moreover,
other kinds of "translation, " such as from technical English, to lay
English are largely ignored and treated as beyond current capabilities.

    To me it appears that if an entity is to be able to translate
languages, it must "understand" languages.   At present, there
are no computer programs which "understand" any but narrowly
circumscribed subsets of English.   This is partly a result of
failure to give adequate accounts of what is to count as "understanding" the language, or perhaps by equating understanding with
the "grasping of meanings" without realizing that such an as-
sertion is for the most part naive philosophy and has little practical significance.   To study understanding, one must study what
it means to communicate and what is the nature of communication between intelligent beings--that is, the human activity which
gives language its meaning.   Prior to the study of computer language
translation, I believe it is necessary to study first the computer
acquisition of language, that is, computer program models of the
acquisition of language.   Linguists are aware of the fact that the
"rules" of language syntax and morphology require extensive analysis on the part of linguists for their discovery, yet with a human
mind, and a few years of speaking and thinking in a language, it is
possible to use the language in almost perfect conformity to these
rules, even without being able to state them.   It seems plausible,
therefore, to combine into one project several different phases of
research--studies in automatic language translation and manipulation;
studies in concept formation, studies in self-organizing problem-

solving systems, and studies in human psychology.

Such a long-range project would be the construction of a computer model patterned after the human mind. Basically, the aim would be psychological--to find the basic mental mechanisms which enable the child to develop more and more complicated structures of patterning and organizing behavior; specifically, to isolate and examine in detail the role of language in facilitating learning and acting. Such a project would unite psychological studies--many "comparative" studies describe human behavior patterns at ages three, five, eight, ten, fifteen, etc., yet fail to reveal the basic underlying seeds of structure present at age three, which are capable, under appropriate environmental training, of producing all of the future behavior patterns.

An analysis of the mind of the three-year old child would be much more valuable for psychology than outlines of behavior patterns at various ages. Here, "analysis of mind" would be more than merely a framework for describing three-year old behavior, but rather one that could be extended to predict the effects of environment on both behavior and future mental state. Only in this way could a computer model of psychological theory of this sort be of particular advantage, since computer languages or pseudo-languages are very convenient ways of formulating expressions of patterns of concepts and manipulation routines, and such programs are also self-testing. A computer program not only expresses the theory, but the program can be "run," thereby testing the theory.

A concentration on psychological organization processes, rather than on the linguistic structure of specific languages, would bear the following long-range fruits:

1. The linguistic utterances of such a computer program would be more "natural," and would tend to follow the idiomatic, free form of the language.

2. The behavior of a computer would improve with experience.

3. Content analysis and language learning would go on at the same time.

4.  The specific rules both for forming utterances, and for under-
    standing content would not have to be pre-programmed, but ra-
    ther would develop within the machine, according to a pattern
    that is permitted to grow.

5.  A "memory dump" of the advanced stages of the machine's
    self-learning will be effectively a set of linguistic rules, and
    a set of psychological patterns, i.e. the linguist will have the
    syntactic and morphological rules for the language developed
    for him, and the psychologist will have a theory of adult cog-
    nitive behavior developed for him. [2]

6.  Training such a program to develop to the status of a profi-
    cient (for whatever practical level is desired) user of the lan-
    guage can be accomplished by any native speaker of the lan-
    guage, just as any parent can teach the child to speak its own
    language, without any knowledge of grammar or rules.

7.  The same basic learning mechanisms for language apply to any
    language, and hence the same program can be used to work
    with any language at all without any program change.

The necessary changes in research attitude basically involve a
switch from study of specific, fully developed processes to a study
of the underlying psychological processes which are involved in creating the mechanisms which lead to the more completely structured
forms of behavior. The long-range applications of such a change
in methodology are extremely vast in scope. The study of self-organizing processes, when combined with the use of the computer
program and high-speed machines, makes it possible to eventually
run many cognitive processes through their "life cycle," dumping
memory from time to time to observe the complexity of the structures which the machines have organized. The use of simulation
makes it possible to control and manipulate the environment to a
much greater extent than it is possible so to do in "real" situations,
and thus it is possible to analyze the effect of environment, other
speakers, and continued use of language; the effect of language on
problem-solving; the environmental situations in which it is possible

for a cognitive system in contact with others to devise its own language; and many other effects.

The results of the quest for a psychological theory of this sort are extremely valuable, both in the pure knowledge attained, and in the uses to which it can be put. The high-speed computer, and the modern symbolic programming languages are invaluable tools in the acceleration of such research. However, there are a number of points well worth bearing in mind in connection with such research, since in the course of reading I was struck by many shortcomings of a dangerous nature therein.

Research in psychological studies of cognitive processes, and in computer simulations, must not fall into the same pitfalls as did research in "brute force" methods. There should be no limits to the extent to which the computer is permitted to develop new behavior rules and program patterns on its own. For example, in the General Problem Solver, the set of "rules" from which the program has to choose is fixed, and pre-set by the programmer, even though the program itself has the choice of which rules it should choose, when, and in what order. How much more flexible the GPS would be if the rules were created by the program itself, and if the list of "relevant" rules were to continue to expand. Having one of the rules entitled, "Add new rule to rule list" could provide for a program with expanding prospects. The present GPS program can learn task behavior, but can never add new dimensions of efficiency to its search for the right solution. On the contrary, human subjects in Jean Piaget's experiments develop new ways of organizing input to their problem-solving task (such as "concrete operations," "formal operations," "reciprocity groups," etc.), using their former methods as "bootstraps," or means toward the creation of the more efficient methods. In short, one should avoid pre-programming any more than just the "seeds" of learning. That task is itself difficult enough, without adding the additional burden of looking for a priori constraints.

The other important point to notice in research is to avoid

the wrong sort of reductionism in trying to simplify the problem in
the early stages.  Virtually all past researchers have made the error of assuming that it is possible to have a self-organizing system
become proficient at a task only through receiving input data associated with the task itself.  When the task is language, though, it is
necessary to have an input system where nonlinguistic environment
can be "perceived" by the computer, and associations can be made
not only with words and sections of the environment, but with changes
in the environment brought about by actions of the computer's own
doing.  It is only in this way that the so-called functional concepts
of language, concepts which don't fall under Earl Hunt's mistaken
characterization of "definitions in symbolic logic, " can be acquired.
For language learning, it is necessary to have an environment which
is in flux, and where the degree and nature of this flux is in part a
function of the computer's own linguistic and nonlinguistic activity.
In this way, it is possible for other than names of sensory events
to become included in the organism's vocabulary.

However, a reductionism is necessary, for otherwise the tasks
involved merely in obtaining input data in a suitable form for processing in concepts become overwhelming--the field of visual pattern
recognition is a vast one in which certain problems have baffled the
experts for years, and in which breakthroughs come slowly.  The
fact that reduction is necessary, however, does not mean that the
important interacting properties of the environment must be sacri-
ficed, reducing to zero the ability of the computer to react to its
environment.  Neither does it mean that specific subsections of the
environment, such as special tasks, should be singled out, since
the learning processes involved in specific task acquisition all come
from basic organizational properties of organisms which originate in
situations far removed from the task to be learned.  We have only
a limited use for programs confined to word-association tasks, etc.,
or for programs which deal with searching for patterns in static environments.  Neither research line leads to the search for the basic
"seeds of self-organization" the discovery of which is the essence

of our research goal.  We should not limit the scope of interaction, or the extent of environment-machine interaction, but merely the complexity of the input sense organ itself.  After all, Helen Keller, with only the sensory material of tactilely perceived object outlines, plus the linguistic input from her tutor, was able to develop all of the cognitive self-organization systems necessary for problem-solving and language learning within her environment.  (Of course, Helen Keller could perceive more than spatial outlines, but I feel that even if her senses were restricted to this extent, she would have been able to develop her intellectual capabilities to the extent which she did.)  The point is that computer designers should search for a particularly adaptable form of sense input, perhaps an artificial combination combining some of the properties of touch and vision--say a sense organ which could perceive spatial outlines of solid objects from a distance.  As long as the crucial element of interaction and flux remains, concept acquisition can still occur.  It can never occur, however, with the type of "reductionist" models that restrict the environment to the extent that no means is provided for the organism to notice interactions among things and interactions between himself and other things.  If one considers the way a simple word, such as "ought," is learned, one will see the necessity of the type of interaction and association which I have been describing.

A recommended set of action routines for a computer learning mechanism to have will be explained in detail in a later paper; however, in brief, they include: Move new sensory information into "direct view"; approach new sense data; store sense data; associate characteristic of sense data with current concept; set up a recognition formula of current sense data pattern, name it and store it; associate current pattern with combinations of previously stored patterns; add new action to this action list; automatically create routine for desired task by "internal feedback" (done in IPL-V by running a "trial" routine under the supervision of a supervisory program, then modifying it in a way determined by the supervisory program, in an effort to bring it closer to the desired goal, then transferring control to the new "trial" routine, etc.)

For example, by means of list structures or by other means, a three-dimensional spatial network can be modeled in computer memory. The machine is able to move in directions corresponding to the three spatial dimensions of its environment, and thereby to bring the "contents" of three spatial positions into view by varying its motions. But it must "learn" the three-dimensional character of its space by acquiring an association between an object perceived in the visual field in a certain way and a certain direction of the "machine's motion" necessary to bring the object into the center of the visual field. Both the visual field and the "objective field" can be represented in storage, and the task simulated is that of sensory coordination and the beginning of the child's creation of the three-dimensional framework in which objects are located. This character of space, and the solidity, permanence, and extension of the concrete object are acquired characteristics and representative of the mechanisms of learning. The shift of the child's concept of an object as vanishable (for example, when it "disappears" behind a closer object) to that of the object as permanent is a change in the categories in which he views the world (or, for a machine, the way the "perceptions" are coded by the learning unit) and in the child's resulting actions (in the machine, what conditions result in which "physical" actions). The changing of either the set of conditions for an action or the nature of the coding of perception is in itself an action, and requires a psychological explanation and a computer model. Thus, the concept of association of stimuli with responses is extended, as a stimulus is now a coding of a perceived situation, not the situation itself, with its infinity of possible "stimulus properties."

The simulation of the physical actions themselves is a relatively simple task. A list-structure containing six-item sublists, each cell of which names the six-item sublists representing "adjacent" boxes in a cubical lattice, can represent the three-dimensional

space.  A more economical solution involves treating points in space
as points in a three-dimensional array.   The physical "actions"
would then consist in copying into a "visual field" array the contents of a specified angular segment of this array.   "Motion" would
consist in sweeping this angular segment over the range of visual
space,  bringing different arrangements of objects into "view. "

It is the psychological processes--especially the learning mechanism's own coding of its perceptions--which is the difficult research problem and is not to be confused with the first.   Setting up
a computerized representation of a meaningful chunk of the real
physical world is a programmer's problem,  with a fairly arbitrary
range of solutions.   Determining the way this in turn is coded by a
perceiving, acting and learning mechanism is very different and requires the cooperation of the psychologist,  with  the  programmer.

Such programs can build up to great complexity and self-organizing power.   It is up to future work to try to formulate the
structure by which such routines can develop enough power to perform a satisfactory duplication of human learning behavior.

Part II

This note reports the existence of two programming packages
which are useful in studying concept learning with the help of a computer.   One program augments the programming language IPL-V by
adding four new J-processes and by improving input-output facility.
The second program makes it possible to use this revised IPL-V
system on both the IBM 7090 and the IBM 7040.

The following discusses the reasons and advantages of these
two programs as tools in the study of concept learning.

A concept learning model is viewed as a computer program
capable of using its ability to find general laws for recognizing patterns of input data.   The model is to be able to form classification
rules by which it can describe relations of objects it "perceives, "
and also relations between its own internal state and the situations
it "perceives. "   A good learning model should take advantage of
past concepts and concept-learning techniques in the process of

learning newer concepts, so that its learning rate "snowballs" as
does a young child's.

This kind of concept learning, combined with the ability to
"cascade" concepts, and devise structured internal representations
of the outside world, is believed by many psychologists to be the
essential characteristic of human psychology by virtue of which the
human intelligence has made such great leaps and bounds over the
behavior of animals, who seem to have more fixed concept acquisition techniques.

A computer program which is to serve as a model must be
capable of possessing all these properties. It must therefore be
able to have well-organized yet manipulatable data storage, easily
augmentable and modifiable. The program must be self-modifying
in a similarly organized way. It should be able to handle large
blocks of data or program routines by specification of merely a
name. It must have all the flexibility of a stored-program computer
language.

IPL-V is a computer language with these properties. Its use
of linked data lists, list structures, and attribute lists make organization very handy. It has "wholistic" properties of being able to
deal with large blocks of data or routines by a single name tying
them together. It has a hierarchical structure of routine execution.
It has complete stored program properties, including crosswise compatibility of data and routines. It has the ability to create or tie together blocks of routines or data with the same facility for both, regardless of size of block. With the new "multiprogramming" feature,
IPL-V can have "internal feedback" organization which I shall discuss later. Moreover, IPL-V is conceptually easy on a programmer, and enables him to view concepts (artificial or human) in a
true "symbolic" manner--i.e. as dualities--both as "things" mani-
pulated in the mind, and as "rules" for manipulating other "things."
In IPL-V, since a symbol can be used for a routine name, it is
quite convenient to represent concepts as routine names, or as
names of lists of routines. Therefore, the name of the concept can

be stored as a piece of data, and then used as a recognition, transformation, or cascading rule, at another time, by merely transferring control to the symbol. IPL-V is also adaptable for computer
simulations, for into it can be inserted special computer-oriented
routines to handle special problems peculiar to computer simulation,
such as input-output mode, tape handling, etc.

Although IPL-V, and (at present) IPL-V alone,[3] met all the
requirements necessitated by the demands of simulation research,
it was, and still is, necessary to make adaptations to IPL-V, to try to
overcome some of its disadvantages. Because of its past history,
and large number of developers in different places, it is somewhat
a "patchy," disheveled language to learn. Although its internal
routines (called "J-processes") are extremely handy, their exact
descriptions don't all follow a simple pattern, and the complement
of processes is badly arranged.

In addition, it appears as if those who devised IPL-V originally did not fully realize the true extent and power of their language,
for they failed to realize that an essential virtue of IPL-V was the
intercompatibility between data and routines, and the extreme power
of being able to construct pieces of routines at object time, as well
as data. It was necessary to remedy this omission by including
four extremely simple J-processes (viz., save and restore P- and
Q- operation code prefixes) to provide for complete isomorphism to
a stored program machine.

In order to introduce IPL-V as a working system, I decided
to incorporate it as a program for the IBM 7040. The creation of
a truly efficient 7040 IPL-V program would involve rewriting the
entire system. However, it was possible, by automatic means (involving FAP's "OPD" and "MACRO" provisions, which can modify
programs in an ordered way at pre-assembly time) to replace all
of the uses of operation codes used on the 7094 but absent on the
7040 by "traps" (skips) to a routine which would interpret such
operation codes. Such a method has the advantages that no extensive rewriting is necessary, that interpretation cycles on the 7040

occur only for instructions of the 7094-but-not-7040 variety and not on all instructions, that any changes to the 7094 version of the IPL-V system automatically become changed in the 7040 version, and that the MACRO package inserted before the FAP deck of the 7094 program to convert it to 7040 can be inserted before any 7094 program to convert it to 7040.   The scientific community interested in the investigation of concept-learning should be able to take advantage of the new system development by using the IBM 7090 as well as the IBM 7040.

This 7094-7040 conversion of IPL-V is a completed program package.   The additional routines for increased compatibility and input-output facility have also been completed.

### Notes

1.   This work was carried out at IBM during the summer of 1963, under support of contract AF19(628)-2752.

2.   The reader may, however, correctly object that while in some trivial sense, any complete description of a device capable of producing representations of human linguistic behavior is a linguistic theory, but of course, realistically, such a description may be in extremely poor form for the purposes of the descriptive linguist, who wants a useful, workable description of a language, not of a language user.

3.   The rapid increase in the numbers and powers of symbol-manipulating languages makes this statement less valid now as a research suggestion than at the original time of writing.   However, the warning against programs that are in essence no more than "algorithms" is still a timely one.

## II D.  Semantic Diversity and a "Growing" Man-Machine Thesaurus [1]

Phyllis Reisner

Abstract:   The diversity in human endeavor is reflected in the "semantic" diversity of a group of Information Retrieval system users. One device for maintaining the intellectual cohesiveness of the group, in spite of this diversity, is the I. R. thesaurus.   However, explicit criteria for construction and monitoring of this device are lacking. This paper describes an attempt to study the overt semantic problem of Information Retrieval.   The eventual aim is to provide such criteria for thesaurus construction and maintenance.   The mechanism for this study is a "growing" thesaurus, which is intended to be both a study tool and, eventually, a useful retrieval device.

### Introduction

This paper is part tutorial, part progress report.   In the tutorial section, the "semantic" problem underlying the information retrieval process is discussed.   In the progress report section, the idea of a "growing" thesaurus as a tool both for studying this problem and perhaps for mitigating it, is presented.   Plans and progress in testing of the idea are discussed.

### Terminology

There is considerable diversity in the interpretation of the following terms.   The meaning we chose for them is therefore indicated.

Semantic:   Pertaining to the meaning of words or word-groups.

Meaning:   The relation between sign (word) and object desig-

117

nated.

Indexing Language:  The set of signs (lexicon) and rules for
interrelating them (grammar) to be used for labelling and requesting
documents in an Information Retrieval System.

Indexing (and Retrieval) Procedure:  The procedure for using
the indexing language to index and retrieve documents.

Indexing (and Retrieval) System:  An indexing language together with the associated indexing and retrieval procedure, and the
user and document populations.

"Near-Synonyms":  In an indexing system, any pair of terms
which can be substituted for each other in indexing documents and/or
in requesting them.   The relation of "near-synonymy" includes
terms related by abbreviation, class inclusion, etc.

"Homographs and Polysemes":  Words with one or more completely or partially different meanings.   We can consider either that
we have one word with different meanings, or two words, spelled
alike, each with its own different meaning.

Thesaurus:  A device indicating the "near-synonyms" in an indexing system.   A thesaurus may indicate that a "near-synonym"
relation holds between a pair of terms, both of which are in the indexing language, or between a pair of terms, one of which is in the
indexing language and one in a natural language.

The Semantic Problem of Information Retrieval

Why is information retrieval a problem?   The answer has to
do, on the surface, with language, and underlying that, with the social entity which creates and uses it.   The reason for the problem
is a very familiar one: the meaning of a word--the relation between
the sign and the object designated--is not a one-to-one relation.

On the one hand, a word can have several meanings: such
words are "homographs" or "polysemes. "[2]   On the other hand,
two words or phrases can have the same, or nearly the same meanings:[2] such words are "near-synonyms. "  For example, this object
I am writing on is: a desk, a piece of furniture, a grey object, a
steel object, a writing aid, a smooth-surfaced object, a disorderly

accumulator of papers and books, etc.  These are, in a sense,
"near-synonyms." A crane can be both a bird and a mechanical
lifting device that resembles one.  These are "homographs."

The homographs in a language cause it to be semantically ambiguous, and the synonyms, to be semantically redundant (in the
usual sense of these terms).  This semantic ambiguity and redundancy of language exists at various linguistic levels: with word endings, with words, with word groups.  (Context, in natural language
prevents a serious problem from arising.)

Whether a language with a given pair of words is semantically
ambiguous and/or redundant depends on the point of view of the
speaker.  For example, in an elementary linguistics textbook, two
languages are described, one which divides the color spectrum into
seven primary colors, the other into two. (Fig. 1)

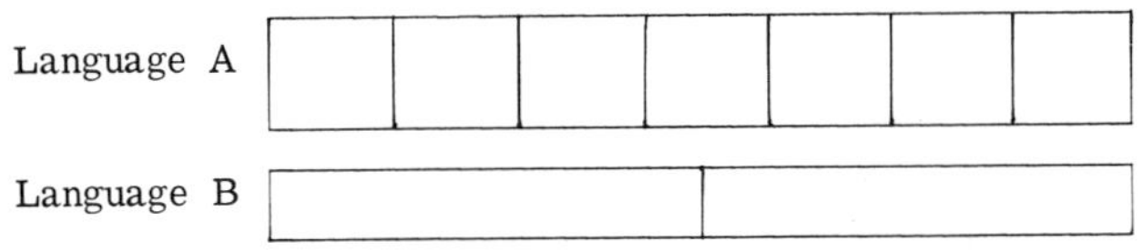

Language A

Language B

Figure 1

Now, it is clear that, from the point of view of the speaker of language A, language B is ambiguous.  (One term in B corresponds to
several in A, i.e. language B contains homographs with respect to
A.)  And, from the point of view of the speaker of language B,
language A is redundant.  (Contains synonyms--several terms in A
where B would use one.)

It is not implausible to assume--and this is to be explored--
that an analogous situation exists within a single language, between
different speakers of the language.  Depending on his own personal
interests, each person, to some extent, views the world differently,
and these differences will be reflected in the semantic interrelations
he chooses (which words are "near-synonyms").  We might, stretch-
ing the analogy somewhat, say that, in the context of information
retrieval, each semantic ideolect is different.  If this is true, then

the implications for information retrieval are clear: the extent of such semantic diversity between users of an information retrieval system determines how close to a perfectly functioning indexing system one can hope to come.

For example, two words, "aircraft" and "airplane," might be considered "near-synonyms" by person a but not by person b. If the indexing language of an IR system treats these as two separate terms, the system will be redundant for person b (two words instead of one). If the language does not treat them as separate terms, the system will be ambiguous for person a (one word, two different meanings). The designer of an indexing language is thus faced with a dilemma. Ideally, he would like to design an indexing language which would be neither semantically ambiguous nor semantically redundant, for ambiguity could cause the user of the system to receive "trash" (he would receive documents labelled with the wrong meaning of the term) and redundancy could cause "miss" (if he searches under one of a pair of synonyms he will miss documents indexed under the other). [3] However, if the language system designer constructs a system for user a, and if the needs and consequent semantic preferences of user b are different from those of a, then necessarily b must suffer.

The Problem of the Indexing Language Designer

The designer of the indexing language for an IR system has thus a twofold problem.

First of all, he must design an indexing language for a potential group of users. But this group is a collection of separate and different individuals. An optimal indexing language for one individual would almost certainly not be optimal for another. If the problem of synonymy (semantic redundancy) is avoided for one, a problem of homography (ambiguity) is necessarily and automatically created for another. Thus the designer must create a kind of compromise language, a stereotype group language to be used as a group tool. He must further, if design of indexing languages is to become a rational and conscious procedure rather than a guessing game, be able to predict how well a system will function (in terms of miss, trash, and work) given a particular compromise language, a given user group, and a given document population.

Except on intuitive grounds, criteria for construction of these compromise languages are lacking.  It is clear that the indexing language should be complete (anything a user or indexer would want to describe could be described), and as semantically nonambiguous and nonredundant as possible.  However, precise and measurable formulations of these admittedly nebulous notions are as yet nonexistent. And of course, prediction of system functioning--perhaps in terms of some kind of semantic diversity (how differently do users semantically interconnect words)--equally so.

Not only are criteria lacking, but methods of constructing these indexing languages are lacking, too, except, once again, on intuitive grounds.  (We limit this discussion to the lexicon and the thesaurus.)  It is clear that, if there is a common core of agreement between users--all users agreeing that certain terms are "synonyms"--then the system designer would want to create a single system term to represent these.  If fewer users would consider the terms synonymous, then the system designer would want to create a cross-referencing structure--or guide to these semantic interrelations--without merging the terms.  But whichever (merging into a controlled vocabulary or cross-referencing) device he chooses to control semantic diversity, how can he tell which terms will be considered synonyms by his users in the first place?  This brings us to the second problem: to what extent can one individual detect the needs of the group entity?

Thesauri, at present, are created by small groups of experts, or by a single expert.  These experts attempt to guess the semantic preferences (which words are "near-synonyms") of potential users. It is not known to what extent such guessing is adequate.  Can any individual, or any small group of individuals, hope to detect the "semantic" preferences of the entire community they represent? Or, will individual bias, or even the largely accidental inclusion of one particular expert, play a decisive role in the thesaurus constructed?  In other words, do we have the same problem in construction of a thesaurus for indexing as we have with the indexing process itself?[4]  Will two different groups of people, in exactly the same situation, build substantially the same thesaurus?  How large a "committee" would we need to build such devices before such uniformity would be manifest?  How effective can the purely guesswork procedures used by these small committees hope to be?  And even once a thesaurus is created, how can changes in user preferences be detected and quickly incorporated in the thesaurus?  In other words, how can the thesaurus be made responsive and sensitive to change?

A "Growing" Thesaurus

To gather data on this semantic problem, and if possible, to alleviate it, the idea of a "growing" thesaurus system is being explored.  This growing thesaurus is based on the general purpose AMNIP system described elsewhere in this report.  The central idea is fundamentally very simple: let the users themselves build the thesaurus in the course of using a man-machine system, use the computer to compile these individually recorded semantic preferences, and let each user choose, from the composite thesaurus, those interrelations that correspond to his own viewpoint.  Thus the computer is to serve as a simple linguistic data gatherer, recording the rare, but hopefully not unique, linguistic "acts" of individual isolated users in a kind of group memory[5] to benefit future ones.  Rather than guessing user semantic preferences, these will be directly recorded.  Continuous updating of the thesaurus is inherent in the system.  In a sense, users will collaborate with each other via the central man-machine computer system.

The AMNIP System

The AMNIP system, as specified by M. Kochen,[3] is an "adaptive" information retrieval system in which the users of the system "train" its memory by entering statements in a semi-formalized "language" through a keyboard console.  The language consists of sentences in the form RLR, e. g. :

|  R  |  L  |  R  |
|-----|-----|-----|
| (Scott) | (is the author of) | (Ivanhoe) |

The memory is conceived as a graph in which the R's correspond to nodes and the L's, the links between the nodes. Programs for this system (by F. Blair and D. Reich) have been written in a modified version of IPL-V and, among other functions, permit a querist to extract from the store lists of sentences or partial sentences of the semiformalized (RLR) language and to perform Boolean operations on these lists. (For a more detailed description, see reference (3) and chapter I A, this report.) The thesaurus system relies heavily, as might be expected, on these extraction, conjunction, negation and disjunction routines.

Function of the Thesaurus System

The user of the Thesaurus System will insert query words into the system via the keyboard console, which will respond with the number of items (e. g. documents, abstracts, etc.) associated with those words. He will then have the option of expanding or limiting his query, or both, on the basis of his own needs and the frequency count displayed to him. The machine will assist him by displaying, on demand, words with which to expand or limit his query. This "man-machine conversation" can be repeated up to the exhaustion of the machine's vocabulary of associations and/or the user's need. Thus the user could insert the additional terms suggested by the machine into the system, find new terms related to those previously suggested, and continue this "path-tracing" query reformulation as long as desired. When the machine's store of associations proves inadequate, the user will enter his own word associations into memory, (e. g. word a is synonymous to word b) leaving a trail for future users of the system, and thus "growing" the thesaurus.

Description of the Thesaurus System

The Thesaurus System has three modes:

1. An Index Use Mode
2. A Thesaurus Lookup Mode

### 3.  A Thesaurus Update Mode

In both the Indexing Mode and the Thesaurus Lookup Mode, the user keys in a question and the machine displays an answer, while in the Thesaurus Update Mode, the user is keying data into the machine. For each of the modes, one or more "links," or "L's" in the RLR statement format specified for the AMNIP system, have been assigned.  Both queries and responses are in this RLR format.

Index Use Mode

The Index Use Mode performs two functions:

1.  Display of a frequency count of the number of items that satisfy a query.

2.  Display of the desired bibliography.

This mode is analogous to normal use of any book-type index, with, however, the possibility of asking complicated queries (difficult if not impossible in book form) and of reformulating a query immediately if the scope of the answer is too large or too small.

The $R_i L_x R_j$ statements for this mode are of the form (word a) (is an index term for) (document b), where $R_i$ is an index term and $R_j$ is a document number.

Thesaurus Lookup Mode

If the frequency count obtained in the Index Mode is not satisfactory, the querist has the option of expanding or of narrowing his query to obtain more or fewer documents.  In the Thesaurus Lookup Mode, he searches for terms with which to do so.  (This mode could be considered the machine's "teaching" mode.)

To expand a query, the user forms a new query consisting of his original term or terms plus the machine-suggested "expander" terms, using the Boolean union operation of the AMNIP system (e.g. everything on abstracts or abstracting).  To limit a query, he constructs a question consisting of his original term (s) plus the machine-suggested "limiter" terms, using the Boolean intersection operation (e.g. everything on both machine and indexing).

For this thesaurus lookup mode, the "R's" of the RLR statements are words or groups of words usable as index terms.  There are several relations, or "L's" between these terms, four to permit expanding a query and one to permit limiting it.[6] The four "expander" L's, linking terms which can be used with the Boolean union routine to obtain a more extensive list of documents are:

1. Synonymy,[7] e. g. (KWIC) (is synonymous with) (CWIC).

2. Spelling or morphological variation, e. g. (abstractor) (is a morphological variant of) (abstracting).

3. Subsumption and supersumption,[8] e. g. (cat) (is included in) (mammal).

4. Unspecified relation. This is a "catch all" link for terms not related as in 1, 2, or 3.

The terms (R's) linked by these L's are obtained in part from the system originator, who builds a small "infant thesaurus" to prime the system, and in part from the users of the system.

There is one "limiter" L, linking terms which can be used, with the Boolean intersection routine, to reduce the number of documents in a response. These terms are obtained, in part from the users, in part from the document store, and are expected to be mainly words with a high frequency of co-occurrence, e. g. (translation) (occurs with) (mechanical). For an example of use, suppose a querist has asked for data on translation of Russian articles and received a count of 500 documents. The machine suggests "mechanical" be used with "translation" to limit the size of the answer. The reformulated query should yield only those documents indexed under both mechanical and translation. Use of this L is clearly not without its dangers (false drops).

Another "limiter" L is contemplated in which the full context (complete phrase, sentence, title, etc.) from which the term was derived will be displayed.

Thesaurus Update Mode

If the Thesaurus Lookup Mode does not provide enough "clues" for query reformulation, the user will enter the Thesaurus Update Mode, inserting RLR sentences into the memory. The machine will check to see whether the suggested words have actually occurred as index terms and will cumulate them for future use.   Except for the fact that data rather than queries are entered into the system, the RLR statements for this mode are the same as for the Thesaurus Lookup Mode.   (This mode could be considered the machine's "learning" mode.)

Data Base

The data base for the current experimental system is derived from "Current Research and Development in Scientific Documentation, No. 10" (10).  The index terms for the Index Mode were obtained by using a concordance program on the subject index to this volume, and then generating RLR statements in the desired format. We considered each project in the NSF volume a "document."
Thus, if the first entry in the subject index were: abstracting, by author, 1.2, we would store:

          (abstracting) (is an index term for) (1.2)

          (author) (is an index term for) (1.2)

Index entries from titles listed in this document have also been prepared.   The reasons for choosing this simplest of the automatic indexing techniques are given in (5).

For the Thesaurus Lookup Mode, a small "infant thesaurus" has been prepared, based on the same data.

Testing

Testing of this idea involves testing of the machinery and programs (search times, etc.), testing of the "man" part of the man-machine interaction process (how hard is it to use, to learn to use, to motivate people to use it) and testing of the underlying assumptions about user semantic behavior.   Due to delays in the implementation of the AMNIP system it was decided to test the "linguistic" portion of the idea independently.

A series of interrelated paper and pencil experiments is, therefore, being run in collaboration with a large scale existing information dissemination system (SDI).

For these experiments, a list of all words used by all querists has been compiled and the frequency of occurrence of each term has been noted.   Each querist in the test is to be presented with a sample of words chosen from his question (profile) and asked to generate "near-synonyms" for each term.   For each term, a composite of the synonyms generated by all users (in the test) for that term will be compiled (the thesaurus will be "grown").   The composite list is then to be returned to the individual user with the request to check those words he considers "near-synonyms."   These checked terms are to be added to his profile and changes in retrieval (miss and trash) noted.

Design and Preliminary Data

   The test population is defined to be the population of word-occurrences, rather than of words.   Thus each occurrence of a word in a query is a test unit.   If we regard the corpus of questions as one text, then the population would be the population of tokens, rather than of types.   Thus, each population unit is, in essence, a word-person pair (a word used in a person's query).

   The first question we wish to ask is:   Do terms re-occur in queries, or is every term in the corpus unique?   (Without such recurrence, the growing thesaurus idea is impossible.)   Examination of the frequency distribution (number of occurrences per query term) has been obtained.   In this corpus, approximately 80 per cent of the tokens occurred more than once.   This permitted us to continue to question II.

   Experiments for question II are now under way.   Here we want to know:   Is a thesaurus constructed in this way necessary at all, or will the user generate all "near-synonyms" himself?   To test this, the number of synonyms originally generated by a user is to be compared with the number he selects from the composite list. The hypothesis to be tested is that the average difference between the number of synonyms obtained by both procedures is zero. Since our population is the population of word-occurrences (tokens) variations in profile size (number of other terms in the query) and in word frequency (number of other occurrences of the term) are to be observed for the tokens sampled.

Question III would involve testing of changes in retrieval but has not yet been initiated.

Data will be gathered on semantic diversity (extent of disagreement between users on which-words-are-synonyms); correlations will be looked for between word-frequency and number of "near-synonyms" of the word; "near-synonym" frequency distribution will be obtained. [9]

## Summary

The growing thesaurus system described represents an attempt to gather data on the underlying semantic problem of IR and, at the same time provide a usable retrieval tool.   At this point, admittedly, we are still groping in an area where precise definition is lacking, data is absent, and problems are not formulated precisely enough to be solved--even to the point of determining to what extent solution is possible.   Thus, while a usable tool is hoped for, the data to be gathered is (to the author) of greater importance.   The eventual goal, of course, is conscious and rational design of indexing languages.   However, this is not yet near.

### Notes

1.   This work was supported in part by the United States Air Force under Contracts AF19(626)-10, AF19(626)-2752, and AF30(602)-3303.   Portions of this paper appeared in "Automation and Scientific Communication," Proceedings of the ADI Annual Meeting, 1963, under the title "Construction of a Growing Thesaurus by Conversational Interaction in a Man-Machine System."

2. This is an oversimplification.   In a sense, every occurrence of a word is different from every other.   And similarity of meaning, too, is a somewhat nebulous notion.

3.   Both of these errors can be avoided, but at the cost of extra work, so our problems are miss, trash, or extra work.

4.   Indexing inconsistency is a major problem in both conventional and nonconventional IR systems.

5.   At this stage, the group language is just the sum total of

the individual ones.

     6.   These distinctions, from the user's viewpoint, are unnecessary.  They were made for: (1) testing the effect of various types of word substitutions on retrieval efficiency (hit rate, etc.), and (2) later restructuring of the memory and partial automation of the lookup procedure to improve operating efficiency (speed and storage).

     7.   Abbreviations and spelling variants are considered synonyms.  "True" synonyms are expected to be quite rare.

     8.   Subsumption and supersumption are considered as one "link" with an "opposite voice" transpose since only one statement is entered as data, the AMNIP system automatically generating an "opposite voice" statement.

     9.   See paper II E for a mathematical technique for comparing thesauri.

## References

(1)   Holm, B. E., and Rasmussen, L. E., "Development of a Technical Thesaurus," American Documentation, July, 1961, pp. 184-189.

(2)   Kehl, W. B., Horty, J. F., et. al., "An Information Retrieval Language for Legal Studies," Communications of the ACM, September, 1961, pp. 380-389.

(3)   Kochen, M., Abraham, C., and Wong, E., "Adaptive Man-Machine Concept-Processing," IBM Final Report on Contract AF19(604)-8446 for Electronics Research Directorate, Air Force Cambridge Research Laboratories, June, 1962.

(4)   Mooers, Calvin N., "The Indexing Language of an Information Retrieval System," Information Retrieval Today, Simonton, Wesley (ed.).  Papers presented at the Institute Conducted by the Library School and the Center for Continuation Study, University of Minnesota, September 19-22, 1962.

(5)   Reisner, P., "Constructing an Adaptive Thesaurus by Man-Machine Interaction."  Final Report on AF19(626)-10, Feb., 1964.

(6)  Sparck-Jones, K., "Mechanised Semantic Classification,"
Proc. of the First International Conference on Machine Translation
of Languages and Applied Language Analysis, London: Her Majesty's
Stationery Office.

(7)  Swanson, D. R., "Searching Natural Language Text by
Computer," Science, October 21, 1960, pp. 1099-1104.

(8)  Vickery, B. C., "Thesaurus--A New Word in Documentation," Journal of Documentation, December, 1960, pp. 181-189.

(9)  "The Philosophy and Guidelines for Revision of the Thesaurus of ASTIA Descriptors," November, 1961.

(10)  "Chemical Engineering Thesaurus," American Institute of
Chemical Engineers, New York, 1961.

(11)  Current Research and Development in Scientific Documentation No. 10, National Science Foundation, NSF-62-20, May, 1962.

II E.    Techniques for Thesaurus Organization and Evaluation[1]

C.  T.  Abraham

Abstract:  For a vocabulary list with synonymy and hierarchy as
the two kinds of pairwise relationships defined between terms, techniques have been found which will (a) group together terms which
are mutually synonymous, (b) provide consistent hierarchies between
terms when they exist and (c) detect inconsistencies in the specification of pairwise hierarchical relationships.   In addition, methods
of comparing two thesauri for consistency using overlapping vocabulary lists have been developed.

An information retrieval thesaurus lists the terms of the system vocabulary and exhibits relationships among these terms.   A-
mong these relationships are the important ones of synonymy and
hierarchy.   Some of these relationships may be hierarchical from
particular points of view but not generally.

The preparation of a thesaurus is a laborious task.   There are
limitations to human ability to compare several terms at the same
time so as to determine whether the terms of a list are all synonyms of one another.   There are, of course, very few true synonyms other than spelling variations or abbreviations, and very few
fundamental hierarchical relationships except in the context of a giv-
en system.

For the purpose of the study of techniques for thesaurus organization, we shall assume that the whole vocabulary list has been
chosen.   Further, we presume that each pair of thesaurus terms
has been examined and the nature of semantic relationships between

these terms, such as synonymy or hierarchy, has been determined. It is quite possible that a small fraction of the total vocabulary has been examined at a time and the exact nature of term-term association has been determined for each term pair in the set.

We wish to model the thesaurus as a graph whose vertices correspond to the terms and whose edges correspond to the term-term semantic associations.  It is obvious that the synonymy relationship is a symmetric pairwise relationship whereas the hierarchical relationship is nonsymmetric.  Thus for each term pair in the thesaurus, there is a corresponding vertex pair in the graph.  If the pairwise relation between two terms is synonymy, this will be indicated by two edges between the vertices with opposite orientations.  The usual hierarchical relationship used in thesauri is the one which specifies that a given index term is more general than another (equivalently, the latter is more specific than the former). In the graph, this will be represented by the two corresponding vertices and an  edge whose orientation is from the vertex corresponding to the more general term to the vertex corresponding to the specific term.  Thus the graph representation of a thesaurus is a large directed graph with multiple edges.

In our attempt to organize thesauri, though the main emphasis has been on the theory and algorithms which affect the organization, the following factors received considerable attention:

1. Ability to deal with large graphs.
2. Ability to optimize the organization of the whole data by optimum organization done in stages, so that the optimality of any particular stage guarantees optimality of succeeding stages.
3. Dynamic reorganization capabilities when new data are added.

By modelling the thesaurus in terms of a directed graph, we hope to achieve the establishment of hierarchies and also the identification of synonym sets using two properties of the graph which are class properties for the set of vertices and the set of edges. The partition of the vertex set into mutually exclusive classes is accomplished by using an equivalence relation on the vertex set.   This

equivalence relation is called cycle-connectivity of the vertices.
Two vertices, A and B, are said to be strongly cycle-connected if
there is a directed path from A to B as well as a directed path
from B to A. (Sometimes this property is referred to as "strong"
connectivity.) For a graph with a large number of vertices, the
partition of the vertices into cycle-connected subsets of vertices can
be accomplished with ease, as shown in an algorithm to be described later. Each subset of the vertex set of the original graph
and the edges connecting vertices in the subset form section graphs.
In graph theory, the section graphs obtained by using the strong cycle-connectivity property are called the leaves of the graph. A further decomposition of the leaf graphs into smaller subgraphs is accomplished by using a class property of the edges of the graph.
This class property of the edges is called strong circuit connectivity which is defined as follows: two edges, say, $e_1$ and $e_2$ of a
graph are said to be strongly circuit connected if there exists a sequence of circuits $C_1$, $C_2$, . . . $C_k$ such that $e_1$ lies in $C_1$ and $e_2$
lies in $C_k$ while any pair of consecutive circuits $C_i$, $C_{i+1}$ have at
least one edge in common. It is obvious that any two edges are
strongly circuit edge connected if they are in the same circuit. The
section graphs obtained by this property are called the lobes of the
leaves of the graph. The following very general properties make
the leaf and lobe decomposition highly desirable and necessary for
thesaurus organization. Without attempting to give proofs we shall
list a few of these properties. They are:

1.  Leaf decomposition partitions the vertices into mutually exclu-
    sive classes.

2.  Any leaf graph is a union of its lobe graphs, i.e. if L is a
    leaf and if $L_i$* are the lobes in the leaf, then:
    $$G(L) = \sum_i G(L_i^*)$$
    where G(A) represents the graph of the vertex set A.

3.  In a directed graph, the separating edges of a leaf are all di-
    rected away from or toward the leaf. (If each leaf is regarded
    as a single vertex, then the resulting graph is a tree.) An edge

e is called a separating edge in the graph G if in the graph H obtained from G by the removal of e, the end points of e are disconnected.

4.   Any two lobes can have at most one vertex in common.

5.   Every vertex of attachment of a lobe graph is a separating vertex.   (In a graph G, a vertex $V_0$ is said to be a separating vertex or cut vertex if there is no proper nonvoid subgraph H with $V_0$ as its only vertex of attachment.   A vertex of attachment of a lobe graph is a vertex it has in common with another lobe graph.

## Algorithms for obtaining consistent hierarchies in a thesaurus

These algorithms have been described in reference 1 as algorithms for obtaining the leaves of a directed graph.   The possibility of dynamic reorganization when new vertices are added to the graph has also been described in that reference.

We shall here only illustrate the use of the algorithm for obtaining leaves to find hierarchies by considering a directed graph with 10 vertices.   The connection matrix N of the graph is as follows:

|       |     | 1 | 2 | 3 | 4 | 5 | 6 | 7 | 8 | 9 | 10 |
|-------|-----|---|---|---|---|---|---|---|---|---|----|
|       | 1   | 0 | 0 | 0 | 1 | 1 | 0 | 0 | 0 | 0 | 1  |
|       | 2   | 1 | 0 | 1 | 1 | 0 | 0 | 0 | 0 | 0 | 0  |
|       | 3   | 0 | 0 | 0 | 1 | 0 | 0 | 1 | 1 | 0 | 0  |
|       | 4   | 1 | 0 | 0 | 0 | 1 | 0 | 1 | 0 | 0 | 0  |
| N =   | 5   | 1 | 0 | 0 | 1 | 0 | 1 | 1 | 0 | 0 | 0  |
|       | 6   | 0 | 0 | 0 | 0 | 0 | 0 | 1 | 0 | 0 | 0  |
|       | 7   | 0 | 0 | 0 | 0 | 0 | 0 | 0 | 0 | 0 | 1  |
|       | 8   | 0 | 0 | 0 | 1 | 0 | 0 | 0 | 0 | 1 | 0  |
|       | 9   | 0 | 1 | 0 | 0 | 0 | 0 | 0 | 0 | 0 | 1  |
|       | 10  | 0 | 0 | 0 | 0 | 0 | 1 | 0 | 0 | 0 | 0  |

The reachability matrix P is given by:

|     | 1 | 2 | 3 | 4 | 5 | 6 | 7 | 8 | 9 | 10 |
|-----|---|---|---|---|---|---|---|---|---|----|
| 1   | 1 | 0 | 0 | 1 | 1 | 1 | 1 | 0 | 0 | 1  |
| 2   | 1 | 1 | 1 | 1 | 1 | 1 | 1 | 1 | 1 | 1  |
| 3   | 1 | 1 | 1 | 1 | 1 | 1 | 1 | 1 | 1 | 1  |
| 4   | 1 | 0 | 0 | 1 | 1 | 1 | 1 | 0 | 0 | 1  |
| 5   | 1 | 0 | 0 | 1 | 1 | 1 | 1 | 0 | 0 | 1  |
| 6   | 0 | 0 | 0 | 0 | 0 | 1 | 1 | 0 | 0 | 1  |
| 7   | 0 | 0 | 0 | 0 | 0 | 1 | 1 | 0 | 0 | 1  |
| 8   | 1 | 1 | 1 | 1 | 1 | 1 | 1 | 1 | 1 | 1  |
| 9   | 1 | 1 | 1 | 1 | 1 | 1 | 1 | 1 | 1 | 1  |
| 10  | 0 | 0 | 0 | 0 | 0 | 1 | 1 | 0 | 0 | 1  |

$P =$ (rows 1–10 as above)

The graph and the reordered matrix are given in Reference 1, Fig. 1 and the matrix preceding it.

Now using the algorithm for obtaining leaves of a directed graph, we obtain the leaves as [1, 4, 5], [6, 7, 10] and [2, 3, 8, 9]. An examination of Figure 1 indicates that the vertex sets which form the first, second and third levels of the tree or hierarchical structural are [2, 3, 8, 9], [1, 4, 5] and [6, 7, 10], respectively. Thus we can form all the consistent hierarchies of vertices since any one vertex of the first level followed by any one vertex of the second level which in turn is followed by any one vertex of the third level will form a consistent hierarchy.

If the length of the shortest path between two vertices is used as a distance measure, it is possible to obtain a ranking among the hierarchical word groups. This shortest distance is indeed a metric. Thus in our illustrative example, for every hierarchical triple, there are three co-ordinates, namely the distance of the first term or vertex from the second, the distance of the first from the third and the distance of the second from the third. Consider the two vertex triples [2, 4, 7] and [3, 5, 6]. The coordinates are [3, 1, 3], respectively. These coordinates could be used as weights for selection purposes.

Algorithms for obtaining synonym lists

Since synonym pairs have connections in both directions, it is evident that such terms can only occur within the leaf graphs obtained by the algorithm for hierarchies. A set of mutually synonymous terms forms a complete graph in the graph theoretic formulation. By a complete graph, we mean a graph in which every vertex is connected back and forth to every other vertex. If pairwise synonymy is the data on the basis of which synonym lists are to be prepared, it is very likely that we might like to examine subgraphs with minimum deficiency of connections for completeness. In either case, we are essentially interested in those subgraphs within the leaf graphs which are richly connected.

As indicated earlier, the next level in the decomposition of the leaf graphs is the formation of lobe graphs for which we use the strong circuit connectivity of the edges. This gives a partition of the edges into mutually exclusive sets. But for each such set, the section graph formed by the edges in the set and their corresponding vertices will have vertices in common with other such section graphs. Thus the sets of terms corresponding to the lobe graphs may overlap. We have indicated that any two such section graphs can have at most one vertex in common. But for purposes of obtaining sets of mutually synonymous terms, these common vertices need not be considered. In terms of the connection matrix N, the vertices which are directly connected back and forth correspond to the nonzero elements of the elementwise product of N and $N^T$ where $N^T$ is the transpose of N. We shall denote this product by $N x N^T$.

Both circuit tracing algorithms described in Reference 1, provide us with a complete list of all the paths between vertices. Within each lobe graph, excluding the overlapping vertices, the vertices and their edges will have to be examined in detail for synonymy sets. The increase in time involved for this is considerably offset by the reduction in the size of the vertex set to be examined. It is evident that the vertices that are candidates for inclusion in a complete subgraph are only those which have circuits of length two. Thus we need consider only those vertices in each lobe graph which satisfy this criterion.   For each such vertex, we form the list of all other vertices which are in its two edge circuits.   Then by intersecting these lists after ordering them on the basis of increasing size of the lists, the complete subgraphs are obtained.

Example:   Thus in the illustrative example used for leaf decomposition, the vertices [1, 4, 5] and [6, 10] form lobes.   In addition, they form complete subgraphs.   Thus, they correspond to two synonym lists.   It is also clear that the vertex sets [2, 3, 9, 8], [1, 4, 5] and [6, 7, 10] are all lobes so that the lobe decomposition will not give any further breakdown beyond the one obtained in leaf decomposition.   Now if we were interested in examining deficiencies for the lobes, we notice that the sets [2, 3, 8, 9], [1, 4, 5], [6, 7, 10] have deficiencies 4, 0 and 2, respectively.   Depending on the specific thesaurus terms, a decision by human judgment to make up the deficiencies or to delete some edges can be made.

So far, we have discussed the organization of a thesaurus with two kinds of pairwise relationship defined for the thesaurus terms. We have described, in detail, the algorithms which lead to the generation of hierarchies and synonym lists in such a thesaurus, which in turn may lead to modifications of the thesaurus by human judgment.   Now we shall discuss some techniques for comparison of thesauri.

Thesaurus comparison

We are interested in detecting differences between a thesaurus which has been constructed for users, and the thesauri which users construct by modification of the initial thesaurus.[2]   In addition, we wish to know whether there is any consistency between the modifications of a thesaurus made by two different users or two thesauri compiled by the same user on different occasions.   For our purpose, a thesaurus consists of a class of sets of words, each set consisting of words which are near synonyms of other words in the set.   Some recent efforts in the construction of thesauri are based on a method of substitution [Ref. 2].   From a pragmatic viewpoint, two words are near synonyms of each other if, in almost all recall-

able linguistic utterances, such as sentences, one word may be substituted for the other without changing the listener's response to the sentence.   We shall first consider only one set of near-synonyms.
There are two distinct situations we wish to consider.

In the first situation, we consider a set of n thesaurus terms which, according to one individual, are near synonyms of one another.   Equivalently, let us assume that this set of n words gives rise to a complete graph with n vertices in which, every time a word is a near synonym of another word, an edge is established between the corresponding vertices.   It is evident that there is an isomorphism between the set of n words which are near synonyms of each other and a complete undirected graph with n vertices.   Now suppose another person is presented the same set of words and asked to decide which of these words are near synonyms of each other.   He may have perfect agreement with the first person, in which case he will also generate the same complete graph.   If, however, he modifies the thesaurus, he will do so by saying that certain pairs of words which were near synonyms according to the first person are not near synonyms.   This is equivalent to deletion of certain edges from the complete graph.   Let us suppose that k edges are removed, which results in m connected components $C_1$, $C_2$, . . . , $C_m$.   If $c_1$, $c_2$, . . ., $c_m$ denote the number of vertices in the components $C_1$, $C_2$, . . . , $C_m$ respectively, then the minimum number of edges to be deleted to get these m components is $\frac{1}{2} (n^2 - \sum_{i=1}^{m} c_i^2)$, according to formula (8); in this case, the components $C_1$, $C_2$, . . . , $C_m$ are each a complete graph.   In particular, if m=2, then the minimal removal is equivalent to the case where the second individual simply wants to use a smaller set of near synonyms but agrees with the first individual's decision of near synonyms.   If, however, the second person removes any more edges, then this is an indication of his bias or his disagreement with the first individual.   Thus in evaluating bias in this situation, the standard for comparison should be the situation where the number of edges removed is minimum.

Here $p = [k - \frac{1}{2}(n^2 - \sum_{i=1}^{2} c_i^2)]$ could be used as a measure

of disagreement between the two persons. If necessary, one could take $p_1$ and $p_2$, which are the number of extra deletions in the components $C_1$ and $C_2$, respectively, and normalize them by the total number of connections in $C_1$ and $C_2$. Thus

$$\frac{1}{2}\left[\frac{P_1}{\binom{c_1}{2}} + \frac{P_2}{\binom{c_2}{2}}\right] = r \tag{A}$$

could be used as a measure of consistency.

On the other hand, if one person is asked to evaluate another person's thesaurus, then he will have to examine every pairwise relation that exists within the set. In this case, we are interested in evaluating the probability that the graph will reduce to m connected components with $c_1, c_2, \ldots c_m$ vertices, respectively, when k edges are removed randomly. The probability $P_n[m, c, h; k]$ for this is given by equation (16) [see Appendix]. Here again the magnitude of the bias can be measured by either

$$p = k - \frac{1}{2}(n^2 - \sum_{i=1}^{m} c_i^2) \quad \text{or by}$$

$$\frac{1}{m}\sum_{i=1}^{m}\frac{h_i}{\binom{c_i}{2}} = r \tag{B}$$

If a more detailed measure of bias is required, then a vector

$$\left[\frac{h_1}{\binom{c_1}{2}}, \frac{h_2}{\binom{c_2}{2}}, \ldots \frac{h_m}{\binom{c_m}{2}}\right] \tag{C}$$

could be used.

Now we consider the case when the thesaurus consists of a set of pairwise near synonyms. It is possible that in the set of all such pairwise synonyms, each term may not be a near synonym of every other term. This means the corresponding graph need not be complete. Thus we are dealing with two individuals who generate two incomplete graphs involving the same number of vertices. Let $G_1$

be the graph corresponding to one thesaurus and $G_2$ the graph corresponding to the other. We wish to compare these two graphs and have a measure of agreement or disagreement between the two. In order to do this, we superpose graph $G_2$ over $G_1$ and call the resulting graph G'. If $A_1$, $A_2$, . . . $A_n$ denote the vertices of the two graphs, then on superposing $G_1$ over $G_2$, if between any two vertices $A_i$ and $A_j$, there is an edge, in $G_1$ as well as in $G_2$, then we assign an edge between $A_i$ and $A_j$ in G'. Similarly if between $A_i$ and $A_j$ there is no edge in $G_1$ as well as in $G_2$, then we assign an edge between $A_i$ and $A_j$ in G'. If on the other hand, either $G_1$ or $G_2$ has the edge between $A_i$ and $A_j$ and the other graph has no edge between $A_i$ and $A_j$ then we leave $A_i$ and $A_j$ unconnected in G'. It is obvious that under this rule G' will be a complete graph if $G_1$ aad $G_2$ are identical. The number of edges required to make G' a complete graph is equal to the number of disagreements between $G_1$ and $G_2$. Thus if there are k disagreements, we wish to know whether these can indicate a significant departure from randomness. If k is less than (n-1), then k removals will leave G' connected and if $k > \binom{n}{2}$ then G' will be always unconnected. However, if $\binom{n-1}{2} \leq k \leq \binom{n}{2}$, then G' will be unconnected with a probability $P_n(k)$ given by equation (15) [see Appendix]. If thesauri are used for information retrieval purposes, then the most relevant quantity to know is $P_n(k)$. When G' becomes unconnected by removal of k edges, the second user selects only a subset of all the linguistic units, such as sentences, in which the thesaurus terms of the first user appear. Thus if G' becomes unconnected more frequently than k random removal of edges predicate, then this information can be used to advantage to reduce the amount of retrieval of statements on file that are irrelevant to the second user. The relationship between bias and trash rate is now clear. Trash rate depends, of course, on variables other than bias as well.

In two of the situations we studied for evaluating bias, both users limited their decisions of near synonymity to the same set of thesaurus terms. It is obvious that this is a stringent limitation.

So we will now consider the case where the number of pairwise near synonyms of the first user is $n_1$ and that of the second user is $n_2$. Let us further assume that there are n pairwise near synonyms which appear in both sets. Let $A_1$, $A_2$, . . . $A_n$ denote the common terms. Let $B_1$, $B_2$, . . . $B_{n'}$, denote the number of terms which are in the first user's thesaurus set which are not in the second user's thesaurus and let $D_1$, $D_2$, . . . $D_{n''}$ be the number of terms in the thesaurus set of the second user which are not in the thesaurus set of the first user. Here evidently $n' = n_1 - n$ and $n'' = n_2 - n$. The bias or disagreement between the two thesaurus sets has to be measured not only by the disagreements concerning the terms $A_1$, $A_2$, . . . $A_n$ but also by the size of the nonoverlapping part of the two sets normalized appropriately. Thus in addition to $P_n$ (k), r, $[h_1', h_2', . . . h_m']$ given by equations (15), (B) and (C) respectively, $\dfrac{n_1 + n_2 - n}{n_1 + n_2}$ should also be used.

So far we have discussed ways of checking consistency or bias for a single set of near synonyms and two overlapping sets of near synonyms. The extension of these procedures to a class of thesaurus sets follows naturally when the same set of pairwise near synonyms are employed by both users. Since in this case the thesauri of both users consist of the same class of sets of pairwise synonyms, for each set bias can be evaluated. If there are N such sets of which M are consistent, then $M/N$ can be used as a measure of agreement between the two thesauri. When M = N, there is complete agreement. In the situation when thesaurus sets of one user are not the same as the thesaurus sets of the other user, evaluating each pair of overlapping sets and proceeding to compute $M/N$ cannot be justified. In this situation, a more complicated analysis has to be performed. A discussion of the analysis will be deferred for a future report.

If the thesaurus sets are large and the thesaurus is used in a man-machine retrieval environment, then the thesaurus of the first user could be regarded as the initial thesaurus in the machine and

any user may start with only a very small fraction of a thesaurus
set in the machine and then he may add a few more terms to the
first set to get a second set and then continue this operation of adding new terms until he is satisfied. [3]   The problems of consistency
between the thesaurus stored in the machine and the thesaurus
formed sequentially by the user can be handled differently.   The
techniques for testing consistency in this situation will be investigated in another paper.

Our main effort in this paper had been the definition of a
problem that arises with the use of thesauri.   We have indicated
ways of evaluating consistency (or bias) between thesauri for certain types of uses of thesauri.   The need to quantify such concepts
as bias since these concepts help to improve quality of retrieval has
been clearly indicated.

## Appendix

Given a complete graph, (i.e. a graph in which every vertex
is connected to every other vertex) we wish to find the probability
that the random removal of a fixed number of edges will make the
graph unconnected.   Further, we would like to compute the probability that such a random removal of edges will result in an unconnected graph with a specified number of connected components each
with a specified number of vertices.   We would like also to know
the probability that this random removal will result in a graph
which will consist of a specified number of connected components,
whose vertices are specified.   We shall indicate the relevance of
these calculations to determining agreement between two or more
different man-made thesauri.

Let $A_1$, $A_2$, . . . $A_n$ be the vertices of a complete graph G
with n vertices.   Let any constant number k of edges be deleted
from G.   Evidently $k \leq \binom{n}{2}$.   Let $P_n(k)$ be the probability that the
removal of k randomly chosen edges will make G unconnected.   Let
$a_n(k)$ be the number of ways of removing k edges which will leave
G unconnected.   Then

$$P_n(k) = \frac{a_n(k)}{\binom{\binom{n}{2}}{k}} \tag{1}$$

where $a_n(k)$ has to be determined.

<u>Solution:</u> It is obvious that it takes at least $(n-1)$ removals to disconnect a single vertex and that G can not be connected if less than $(n-1)$ connections remain after removing k edges.

Thus
$$a_n(k) = \begin{cases} 0 & \text{if } 0 \le k \le (n-2) \\ \text{to be determined if } (n-1) \le k \le \binom{n-1}{2} \\ \binom{\binom{n}{2}}{k} & \text{if } \binom{n-1}{2} < k \le \binom{n}{2} \end{cases} \tag{2}$$

Suppose that the removal of k edges disconnects G into m parts $C_1, C_2, \ldots C_m$ where each $C_i$ is connected and has $c_1, c_2, \ldots c_m$ vertices, respectively.  Then

$$\sum_{i=1}^{m} c_i = n \quad \text{and} \quad c_i \ge 1 \quad i=1, 2, \ldots m \tag{3}$$

For each integer $k \, \rangle \, (n-1) < k \le \binom{n-1}{2}$ and integer $m(2 \le m \le n)$ write

$$A_n(k;m) = [(c_1, c_2, \ldots c_m) ] \subset [\text{collection of m-partitions of n}]$$

Let $a_n(k, c; m)$ = number of sets of k edges whose removal will partition G into $[c_i]\sum_{i=1}^{m} = c$

Let $M_n(k)$ = maximum number of disconnected components that removing k edges from G can produce.  Then, we have

$$A_n(k) = \sum_{m=2}^{M_n(k)} \sum_{c \, \epsilon \, A_n(k;m)} a_n(k,a;m), \quad (n-1) \le k \le \binom{n-1}{2} \tag{4}$$

where $M_n(k)$, $A_n(k;m)$ and $a_n(k,a;m)$ are to be determined.  We will first show that $M_n(k)$ is the largest integer preceding

$$\left\{ \frac{2n+1 - \sqrt{8[\binom{n}{2} -k] + 1}}{2} \right\} \tag{5}$$

To see this, observe that one obtains the maximum number of partitions by repeatedly setting $c_i = 1$ until the remaining available removals can no longer disconnect a single point. The number of removals required to produce an m-partition by this process is simply $(n-1) + (n-2) + \ldots + (n-m+1) = n(m-1) - \dfrac{m(m-1)}{2}$ so that for any m and k we have the inequality

$$\frac{(2n-m)\,(m-1)}{2} \;\leq\; k \tag{6}$$

Thus $M_n(k)$ is the largest integer $m$ satisfying

$$m^2 - (2n+1)\,m + 2\,(m+k) \;\geq\; 0 \tag{7}$$

so that

$$M_n(k) = \left[ \frac{2n+1 - \sqrt{8\,[\,\binom{n}{2} - k\,] + 1}}{2} \right]$$

Let $a_n(c;m)$ = the least number of removals required to produce an m-partition with specified $c = [c_1,\ c_2,\ \ldots c_m]$ numbers of vertices in each component. Then

$$a_n(c;m) \;=\; \frac{n^2 - \sum_{i=1}^{m} c_i^2}{2} \tag{8}$$

To see this, we may count

$$a_n(c;m) = c_1\,(n-c_1) + c_2\,[n-(c_1 + c_2)\,] + \ldots$$

$$+ c_{(m-1)}\left[\, n - \sum_{i=1}^{m-1} c_i \,\right]$$

$$= n \sum_{i=1}^{m-1} c_i - \sum_{i=1}^{m-1} c_i \sum_{j=1}^{i} c_j$$

$$= n \sum_{i=1}^{m} c_i - n\,c_m - \sum_{i=1}^{m} c_i \sum_{j=1}^{i} c_j + c_m \sum_{j=1}^{m} c_j$$

$$= n^2 - \sum_{i=1}^{m} c_i^2 - \sum_{i=1}^{m} \sum_{j=1}^{i-1} c_i\,c_j$$

Therefore,

$$2\,a_n(c;m) = n^2 - \sum_{i=1}^{m} c_i^2 + \left[\, n^2 - \sum_{i=1}^{m} c_i^2 - 2\sum_{i \leq j} c_i\,c_j \,\right]$$

$$= n^2 - \sum_{i=1}^{m} c_i^2 + \left[\, n^2 - \left(\sum_{i=1}^{m} c_i\right)^2 \,\right]$$

$$= n^2 - \sum_{i=1}^{m} c_i^2$$

Another way of seeing this is as follows: Express the connections in the form of a symmetric matrix $\| \delta_{ij} \|$ where

$$\delta_{ij} = \delta_{ji} = \begin{cases} 0 & \text{if the edge connecting } A_i \text{ and } A_j \text{ is removed} \\ 1 & \text{if the edge connecting } A_i \text{ and } A_j \text{ is not removed} \end{cases}$$

and the rows and columns are arranged so that the vertices of a connected component appear as a block on the margins:

| | $c_1$ | | | $c_2$ | | | | | $c_m$ | | | |
|---|---|---|---|---|---|---|---|---|---|---|---|---|
| | $A_1$ | $A_2$ | $A_3$ | $A_4$ | $A_5$ | $A_6$ | $A_7$ | ......... | $A_{n-2}$ | $A_{n-1}$ | $A_n$ | |
| | 1 | 1 | 0 | 0 | 0 | 0 | 0 | 0 | 0 | 0 | 0 | $A_1$ |
| | 0 | 1 | 1 | | | | | | | | | $A_2$ |
| | 1 | 0 | 1 | 0 | 0 | 0 | 0 | 0 | | | | $A_3$ |
| | 0 | 0 | 0 | 1 | 1 | 0 | 1 | 0 | | | | $A_4$ |
| | | 0 | | 1 | 1 | 1 | 1 | 0 | | | | $A_5$ |
| | | 0 | | 0 | 0 | 1 | 1 | 0 | | | | $A_6$ |
| | | 0 | | 1 | 0 | 0 | 1 | 0 | | | 0 | $A_7$ |
| | 0 | | | 0 | 0 | 0 | 0 | | | | | |
| | 0 | | | | | | 0 | 1 | 1 | 0 | | $A_{n-2}$ |
| | | | | | | | 0 | 0 | 1 | 1 | | $A_{n-1}$ |
| | 0 | | 0 | | | | 0 | 1 | 0 | 1 | | $A_n$ |

It is evident that $2\, a_n(c;m)$ is the number of elements outside the nonzero submatrices on the diagonal. For a computer algorithm to accomplish this diagonalization, see Reference 1.

Now we define $A_n(k;m)$. Fixing $c = [c_1,\ c_2,\ \ldots\ c_m]$ and $m$, there are $[k - a_n(c;m)]$ removals to be made within the partitions in

such a way that each partition $C_i$ remains connected.   Thus c must
satisfy

$$k - a_n(c;m) \geq 0 \tag{9}$$

Hence

$$A_n(k;m) = [ (c_1, \ c_2, \ . \ . \ . \ c_m) \ ]; \ \sum_{i=1}^{m} c_i = n, \quad c_i^2 \geq n^2 - 2k$$

$$c_i \text{ are positive integers.} \tag{10}$$

The function $a_n(k, \ c;m)$ can now be defined.   Let $h_i$ = number
of edges removed from the connected component $C_i$.   Then $h_i$ are
integers satisfying

$$\left. \begin{array}{l} 0 \ \leq h_i \ \leq (\begin{smallmatrix} c_i-1 \\ 2 \end{smallmatrix}) \text{ and} \\[3mm] 0 \ \leq \sum_{i=1}^{m} h_i = k - \dfrac{n^2 - \sum_{i=1}^{m} c_i}{2} \end{array} \right\} \tag{11}$$

The possible choices for $h = (h_1, \ h_2, \ . \ . \ . \ h_m)$ are in the set of ordered m-tuples

$$H_n(k, a;m) = [ (h_1, h_2, \ . \ . \ . \ h_m) \ ]; \ \sum_{i=1}^{m} h_i = k - \dfrac{n^2 - \sum_{i=1}^{m} c_i^2}{2}$$

$$\text{and } 0 \ \leq h_i \ \leq (\begin{smallmatrix} c_i-1 \\ 2 \end{smallmatrix}) \tag{12}$$

Now there are

$$\binom{\binom{c_i}{2}}{h_i} - a_{c_i}(h_i) \tag{13}$$

ways of choosing $h_i$ edges from the component $C_i$ and there are

$$\left( \begin{array}{c} n - \sum_{j=1}^{i-1} c_j \\ c_i \end{array} \right)$$

ways of selecting $C_i$ with fixed m and fixed $c = (c_1, c_2, \ . \ . \ . \ c_m)$
(Here we assume that the selection is done in the order of sub-

scripts.) Thus for fixed $k$, $m$, $c_i$, $h_i$, there are

$$\prod_{\substack{i=1 \\ i=1}}^{m} \left\{ \left( \binom{c_i}{2} \atop h_i \right) - a_{c_i}(h_i) \right\} \frac{m-1}{\prod_{i=1}} \left( n - \sum_{j=1}^{i-1} c_j \atop c_i \right)$$

$$= n! \prod_{i=1}^{m} \left\{ \frac{\left( \binom{c_i}{2} \atop h_i \right) - a_{c_i}(h_i)}{c_i !} \right\} \tag{14}$$

ways of making removals.

Thus $a_n(k, a; m)$ is the sum over $h \in H_n(k, a; m)$ of the expression (14). Using equations (8) to (14), we get

$$P_n(k) = \begin{cases} 0 \text{ if } 0 \le k \le n-1 \\[2mm] \dfrac{n!}{\binom{n}{2}} \sum_{m=2}^{M_n(k)} c\epsilon A_n(k;m) \sum_{h\epsilon H_n(k, c;m)} \dfrac{\prod_{i=1}^{m} \left\{ \left( \binom{c_i}{2} \atop h_i \right) - a_{c_i}(h_i) \right\}}{c_i !} \\[5mm] \qquad\qquad\qquad \text{if } n-1 \le k \le \binom{n-1}{2} \\[3mm] 1 \text{ if } \binom{n-1}{2} < k \le \binom{n}{2} \end{cases} \tag{15}$$

where $M_n(k)$, $A_n(k;m)$ are as previously defined.

We give below a tabulation of some of the values of $a_n(k)$.

Values of $a_n(k)$

| k \ n | 1 | 2 | 3 | 4 | 5 | 6 | 7 . . . . . . . . . . . . . . n |
|---|---|---|---|---|---|---|---|
| 0 | 0 | 0 | 0 | 0 | 0 | 0 | 0                    0 |
| 1 | 0 | 1 | 0 | 0 | 0 | 0 | 0                    0 |
| 2 | 0 | 0 | 3 | 0 | 0 | 0 | 0                    0 |
| 3 | 0 | 0 | 1 | 4 | 0 | 0 | 0 |
| 4 | 0 | 0 | 0 | 15 | 5 | 0 | 0 |
| 5 | 0 | | | 6 | 30 | 6 | 0 |
| 6 | 0 | | | 1 | 85 | 60 | 7 |

| k \ n | 1 | 2 | 3 | 4 | 5 | 6 | 7 ............... n |
|---|---|---|---|---|---|---|---|
| [continued] | | | | | | | |
| 7 | | | | 0 | 120 | 270 | 150 |
| 8 | | | | | 45 | 735 | 735 |
| 9 | | | | | 10 | 1370 | 3183 |
| 10 | | | | | 1 | 1857 | 9570 |
| 11 | | | | | 0 | 1365 | |
| 12 | | | | | | 455 | |
| 13 | | | | | | 105 | |
| 14 | | | | | | 15 | |
| 15 | | | | | | 1 | |
| 16 | | | | | | | $\left( \binom{7}{2} \atop k \right)$ |
| . | | | | | | | |
| . | | | | | | | |
| . | | | | | | | |
| 21 | 0 | | | | | 0 | 0 |

Equation (15) gives the probability that the random removal of k edges will make the complete graph G with n vertices unconnected. Expression (14) gives the number of ways of removing k edges such that the graph G becomes m connected components with specified number of vertices $[ c_i ]_{i=1}^{m}$ and specified number of edges $[ \binom{c_i}{2} - h_i ]_{i=1}^{m}$. Thus the probability $P_n[m, c, h; k]$ that the random removal of k edges $[n-1 < k \leq \binom{n-1}{2}]$ will result in m components $C_1, C_2, \ldots C_m$ with $c_1, c_2, \ldots c_m$ vertices and $[\binom{c_1}{2} - h_1 ]$, $[\binom{c_2}{2} - h_2], \ldots [\binom{c_m}{2} - h_m]$ edges respectively will be given by

$$P_n [m, c, h; k] = \frac{n! \prod_{i=1}^{m} \left\{ \dfrac{\left( \dbinom{c_i}{2} \atop h_i \right) - a_{c_i}(h_i)}{c_i !} \right\}}{\left( \binom{n}{2} \atop k \right)} \qquad (16)$$

It is also evident that if in addition to fixing m, $\left[c_i\right]_{i=1}^{m}$ ,

$\left[h_i\right]_{i=1}^{m}$ and k, the specific edges in each $C_i$ are also given, then the conditional probability for obtaining m such components is simply

$$\frac{c_i\ !}{n\ !\ \sum_{i=1}^{m}\ \left[\ \binom{\binom{c_i}{2}}{h_i}\ -a_{c_i}\ (h_i)\ \right]} \tag{17}$$

and the unconditional probability for a specific configuration is obviously

$$\frac{1}{\binom{\binom{n}{2}}{k}} \tag{18}$$

It will be interesting to calculate these probabilities for a graph which is not complete initially.

### Notes

1.  The research reported in this paper was sponsored (or sponsored in part) by the Air Force Cambridge Research Laboratories, Office of Aerospace Research, under Contract AF19(628)-2752.

2.  For a more complete discussion of how and why thesauri should be modified, see paper IV A in this volume.

3.  This paper should be read as companion to paper II D in this volume.

### References

(1)  Abraham, C. T., "Graph Theoretic Techniques for the Organization of Linked Data," Final Report on Contract AF19(628)-2752.  AFCRL Cambridge Massachusetts, 1963.

(2)  Sparck-Jones, K., "Mechanized Semantic Classification." Proceedings of the First International Conference on Machine Translation of Languages and Applied Language Analysis.  London, Her

Majesty's Stationary Office.

(3)   Harary, F., "A Graph Theoretic Method for the Complete Reduction of a Matrix with a View Towards Finding its Eigenvalues." Journal of ACM, 1958.

(4)   Ore, O., Theory of Graphs.   American Mathematical Society Colloquium Publication, Vol. XXXVIII.   1962.

(5)   Hall, Marshall, "An Algorithm for Distinct Representatives," American Mathematics Monthly, pp. 716-717, 1956.

III.   The Storage/Recall Subsystem

This subsystem of an information system has the function of
storage and recall.   While this function cannot be considered completely independently of comprehension--the function of the knowledge
subsystem--it is plausible that memory without understanding and
understanding without memory is to some extent possible.

Concrete embodiments of this subsystem that are most in evidence are exemplified by the large libraries and their catalogs,  by
personal libraries and reprint collections,  as well as other collections of documented facts, findings, judgments, discussions, etc.,
which are worth preserving.

Three types  of use of an information storage system should
be distinguished:

A.   Requests for specific documents which the querist knows to exist.   This is analogous to recall of items in memory.   The over-
whelming fraction of requests in a library are now of this type.
The response called for is confirmation that the item is in the collection and a report on its availability.

B.   Requests for documents, both those known to exist and others,
which pertain to a given topic.   This is the kind of query serviced
by the special, mission-oriented information centers rather than by
the discipline-oriented libraries.   To call this search would be more
appropriate than calling it "recall."

C.   Requests, not for information that can be directly looked up by
literature search or recall, but which can be inferred from such information.   This is akin to search and/or problem-solving.   The
storage/recall subsystem and the processing system are very closely
interrelated.   It is, however, useful to distinguish between process-

ing with a minimum of recall. For example, to answer "what is 12x25?" requires recall only of $12/4 = 3$ and "multiplication by 25 is like division by 4." Recall with a minimum of processing is illustrated by recalling that $12/3 = 4$, which requires only the association "$3x4 = 12$."

Storage and recall of documented knowledge involves organization of such knowledge into orderly patterns for the various purposes to be served: detection of unwanted gaps, redundancies, contradictions in the corpus of documented knowledge; bringing needed knowledge to bear on solution of urgent, real problems; anticipation of future problems and preparation for attacking them. Only when it is recorded in organized form can knowledge be useful; if disorganized, as in a random collection of facts (or some arbitrarily ordered collection of facts or document, e.g. by date of documentation) it may mislead more than help.

Textbooks, and, par excellence, encyclopedias, are man's best attempts at recording documented knowledge in organized form. They are the results of his greatest effort to assimilate, integrate, tie together knowledge into a coherent whole sufficiently fast to keep pace with the growth of newly generated knowledge. There are indications that the development and maintenance of a truly comprehensive, frequently updated, semi-automated encyclopedia service is both technologically feasible and would fill an important latent need. Indeed, we feel that the need for a system which helps us to assimilate the vast quantities of knowledge available to us through existing search tools is far greater than the need for more effective search tools. This idea is described more fully in III A by Bohnert and Kochen.

Encyclopedia articles and texts are products of a synthesis of material found in books, articles, reviews and other forms of documented information. One of the fundamental bonds in terms of which the organization of documented knowledge is made evident is that of a citation. It is the explicit link between a document and its predecessors which contributed to its synthesis. (There are, of course, the equally important implicit links, such as the use in a

20th century paper of a well-known theorem discovered by the ancient Greeks, without an explicit citation to any paper proving that theorem; but such implicit links are more difficult to deal with.)

The network of all documents interconnected by such citation bonds is an excellent device for studying the organization of recorded, documented knowledge. The use of certain of the properties of this network could also be a useful, practical aid in recall. By tracing citation trails, starting with some document known to be of interest, and abandoning any trail at a document of no interest, the querist may be led to very many other documents of interest he had not suspected to exist. A planned experiment to demonstrate how the use of such a procedure might improve the use of a document storage/recall system is described in paper III B appended to this chapter. The implementation and use of a citation index based on a time-shared computer system could not only facilitate investigations into the structure of the storage/recall subsystem, but provide improved literature search tools as well.

The desire to minimize mean search time and total storage requirements in storage and recall, as for instance in tracing citation trails, is intimately related to the assignment of stored records to memory locations. One approach to this question of optimal file organization is to explore the applicability of the theory of percolation processes. At first, this was felt to be promising, but further analysis showed that the methods of percolation processes were severely limited for the problems encountered here. Some of these considerations are discussed in paper III C. This paper also describes some experiments for evaluating certain critical parameters of the file. Such results may find applications to the optimal design of computer-based aids to storage and searching, but they also illustrate the important role of graph-theoretic ideas in the intellectual structure of an information system.

Another important procedure is to consider records (e. g. documented knowledge) as represented by a keyword and several secondary identifiers, and to devise procedures for assigning addresses to such keywords and secondary identifiers when recording.  Suppose, for example, that a keyword is 10 characters or 60 bits long.  Very few of the $2^{60}$ possible 60-bit strings are actual keywords.  Yet their number may be large, say 10,000.  In general, it is not possible to devise an algorithm which will, for every 60-bit string corresponding to an English word, assign a unique address among the 10,000.  Generally, a 60-bit string is mapped into an address into which many other 60-bit strings may also be mapped.  Stored in that address is a string of 60-bit words, all of which have to be matched against a search term at query time.

Some important new ideas and results on such problems are reported in papers III D and III E.  These results are but first steps toward the situation when the successive terms in a sequence of query terms must be considered dependent rather than statistically independent, where the frequency with which various keys occur in queries is not the same for all terms, and where the reassignment of addresses to keywords is done on a continual basis as the file grows.

An entirely different approach to the same question is the use of "associative," "content-addressable," "catalog" or "tag" memories as there are variously called.  In a sense these are hardware implementations of what file addressing techniques attempt to do by programs.  When to freeze a lookup/recall procedure into hardware, provide a programming package, use microprogramming is still subject to investigation.  A briefly summarized but fairly comprehensive, critical survey of the state of the art in "associative" memories as the type best suited for the kind of processing involved here is described in paper III F.

Paper III D, by Frazer, investigates the general problem of constructing, by a combination of programmed search techniques and hardware, an image of the data in the file.  This problem includes the card-catalogue approach, the coordinate indexing approach, as well as the approach using associative memories.  The particular idea in this paper is to explore the possibility of forming a hardware image of data in the file as a logical combination of adjacent data words in the file.   In search, the hardware image is searched first and the list of records so retrieved is thus searched.   An increase by a factor of $\frac{n+1}{n}$ hardware results in an increase by a factor of n in search efficiency for the case in which the descriptor words are distributed randomly over the records.

The four papers on file organization, III C, III D, III E, III F, all pertain to questions of encoding records of information so as to maximize efficiency of storage and search, and specifying strategies for storage and search.   When files are very large, these questions are very important for the design of automated aids to storage and recall.

Insofar as paper II D is a quantitative analysis of storage and search it also belongs with this group of papers.   It was included in the chapter on "information science" to illustrate the kind of research that can and should, in the author's view, be stressed to move toward information science.   Paper III D, for example, could equally well have been chosen for this purpose.

On the other hand, paper III A would more appropriately fit into chapter II.   The reason for including it here is that the concept of an encyclopedia is so central to organized storage of information. Most of the technical papers could be considered specific studies of various detailed aspects of attempts to create such an encyclopedia-based system.

III A.   The Automated Multilevel Encyclopedia as a New Mode of
Scientific Communication[1]

H.  Bohnert

and

M.  Kochen

Abstract:   The computer revolution encourages reexamination of the
encyclopedia concept.   No longer confined to book form, it can be
enlarged to embrace several levels of condensation from a library-
sized store, with updating, text preparation, access, and distribution
handled differently at different levels, partially replacing, at base
level, the journal system.

Most of the "information crisis" literature stresses the difficulty of access to the growing flood of technical literature.   But the
problem that more often confronts the researcher is that of absorbing the mountain of material already accessible.   An increase in ac-
cessibility without a corresponding increase in human assimilation
rate will be self-defeating.   Short of discovering a harmless super-
benzedrine, the only way such a thing can be done is by making better use of the natural rate.   At present this natural rate is waste-
fully expended, not only in literature search (the target of most current documentation efforts), but also in wading through series of
"self-contained," hence highly overlapping but terminologically idio-
syncratic, articles in order to build a consistent, single picture of
the state of development in a given field.   Often one needs to know
only a central idea, result, theorem, or the method employed, with
bibliographical information for later reference, but finds that a short

course in unneeded detail is required to get to it.   At such times,
one thinks fondly of the occasions when a quick look-up in an encyclopedia or handbook has yielded an authoritative summary, but reflects with resignation that encyclopedias are necessarily limited in
size, directed to the general reader rather than the researcher, and
usually more than ten years out of date.   So they have been.   The
use of computers, however, can change that.   In fact, a reexamination of the encyclopedia concept in the light of current technology
suggests that it may serve as the basis for a new system of scientific communication, taking in and integrating the latest advances,
giving out what information is required, not only as much but also
as little, tutorial or technical, elementary or advanced.

Since technology offers remote access, it follows that location,
and hence size, need no longer be limitations.   The master encyclopedia need exist at only one center.   It could have a word content equal to thousands of volumes, though of course it need have no
resemblance to book form.   In fact, it probably would not have any
one form.   Technology offers a broad variety of access and storage
modes, so that less frequently used, or less valuable, records
could be stored more cheaply at the cost of slower access.

The master encyclopedia, then, would be an integrated system
of storage systems, discs, tapes, cores, cards, microforms.   It
would contain not only summaries, but articles on many levels of
detail, maps, diagrams, tables, and finally the masses of basic
data of science, law, business, medicine, and government.   It could
approach, ultimately, a representation of the total state of knowledge, rather than provide a reference summary only.

The summary concept is essential to the encyclopedia concept,
however, and technology can expand it to embrace several levels of
condensation.   One volume and thirty-volume abridgements could be
published periodically from the steadily updated master store.
These could be on cheap paper, and be replaced periodically for
each subscriber.   A subscriber could also call for further collections of articles compiled automatically from the master encyclope-

dia, according to his fields of interest.

When the need for fuller or more current information cannot be met by these abridgements, the master encyclopedia could be interrogated from remote consoles, perhaps producing printed copies at that point.

More generally, this system, with its variety of access-currency-detail trade-offs, should provide the effect of a variable-focus, or "zoomar" lens on reality, from the big picture to the ultra-microscopic. Beginning with the briefest summaries, the researcher should be able to follow down more and more detailed or advanced discussions, culminating, when desired, in the most basic data available to society; laboratory reports, patents, case histories, and the like.

It should also permit him to study his way into new disciplines, indicating prerequisites, suggesting course of study, perhaps, through teaching machine consoles, providing detailed instruction.

In the case of formatted material, the system might act as direct question answerer, not only through table-look-up, but also through conditional searches, through computation, summing, correlating, finding maxima, interpolating, or performing symbolic logic deductions from sets of rules, definitions, etc.

Obsolescence, the chronic defect of encyclopedias in the past, can be overcome once the basic text to be updated exists only at one location and not in printed form. A form of storage which permits any number of insertions and deletions yet can serve as a master program for the printing of inexpensive condensations is within the technological horizon.

Updating the master store would require the collaboration of an editorial and technical staff at the center and the scientific community, which would provide the authors and referees. In fact, once the system were established, it seems reasonable to expect that new research results would be presented as amendments to the encyclopedia rather than as separate, quasi-independent articles with their varying recapitulations of background that waste the time of both writer and reader.   Amendments could vary from changing a digit to recasting the whole treatment of a subject.   Refinements too slight to be worth an article, but cumulatively valuable, would accrue because of their ease of entry.   The encyclopedia by its very claim to currency would constitute a challenge to every scientific reader that would guarantee strong corrective feedback.   To be sure, since the encyclopedia would come to have great authority, pedagogical and scientific dispute would arise on points which would be passed over in any one textbook.   It is likely that the referee system would become more elaborate than at present, with provision for appeals, hearings, special panels, and the like.   But such confrontations may prove of value, and indeed accelerate research, if the referee system is designed so that decisions are based, as much as possible, on reason and evidence, and if a policy of fairly representing opposing points of view in the encyclopedia itself is adopted.

H. G. Wells portrays his famous "World Brain" as still consisting of a set of volumes, but the present conception seems in keeping with his intentions.   Watson Davis has remarked on the curious lack of follow-up among documentalists of this fundamental aspiration.   If we are interested in coping with the flood of information rather than just the flood of documents, this new, large "document," and the system to manage it, must be added to the flood.

## Notes

1.  This paper was published in Automation and Scientific Communication, Proceedings of the American Documentation Institute, 28th Annual Meeting, 1963, Vol. 2, pp. 269-270.

## References

Davis, W., Documentation Unfinished.   Statement at the 25th Anniversary Meeting of ADI, Hollywood-by-the-Sea, Florida, December 12, 1962.

Wells, H. G., World Brain, Doubleday Co., Garden City, N.Y. 1938.

Library of Congress, <u>Automation and the Library of Congress,</u>
Washington, 1963, King, G. W.; Edmundson, H. P.; Flood, M. M.;
Kochen, M.; Libby, R. L.; Swanson, D. R.; Wylly, A.

III B.  Pre-Test and Potential of a Machine-Stored
Patent Citation Index[1]

Phyllis Reisner

Abstract:  A citation index to patent literature has been constructed
for use in an experimental man-machine system.   The intended application of this system to citation indexing is described.   Results
of a paper-and-pencil pre-test of the utility of the citation index are
given and some problems in the evaluation of citation indexing are
discussed.

Introduction

Citation indexes are receiving increasing attention as bibliographic aids and as sociometric tools.   As sociometric tools, they
are being used to explore the flow of information across national
boundaries and from pure to applied fields.   They have also been
considered for determining the structure of a field of knowledge,
and for evaluating documents or authors.   Several large scale and
many small scale citation indexes are now being constructed as bibliographic tools--usually for manual use in bound-book form.   If the
notion of a citation index proves to be as fruitful as expected, a
man-machine system to permit complex searches through the citation net could extend the versatility of the index.   Therefore, a
partial citation index to patents has been prepared for use in a general purpose man-machine system (AMNIPS) described elsewhere in
this report.   A small preliminary paper-and-pencil test of the utility of this citation index has been performed while awaiting the
availability of the experimental system for large-scale testing.

161

Problems in Information Science

## Data Base

The data base is now composed of approximately 36,000 patent disclosure numbers, 6,000 original, or citing patents and 30,000 references, or cited patents. Figures given in this section are based on a smaller data base, 3,000 original and 15,000 cited patents. This data base has recently been doubled to the 36,000 figure.

The 3,000 original patents[2] were selected from the Patent Gazette from June, 1959 to June, 1960, with the selection criterion "possible interest ot IBM patent attorneys." The patent disclosure numbers for both the original and the cited patents are available, together with an IBM Classification code assigned to the original patent.

Some properties of the data base are given below:

### Size of Collection

| | | |
|---|---|---|
| 1. | Number of original (citing) patents: | 3,250 |
| 2. | Total number of references (cited patents): | 15,250 |
| 3. | Number of different references: | 9,870 |

### Reference Patent Frequency Distribution (Number of reference patents per original patent)

1. Mean: 4.83 (standard deviation 1.28)
2. Maximum: 30

### Citation Patent Frequency Distribution (Number of citations from reference patents to original patents)

| | | |
|---|---|---|
| 1. | Mean: | 1.5 |
| 2. | Maximum: | 40 |
| 3. | Percentage cited once: | 50 per cent |
| | cited five times or less: | 83 per cent |

### Time Spread

Although the original patents were selected from one year only, the cited patents extend from the early nineteenth century through June, 1960.

### "Inbreeding"

The set of cited and the set of citing patents were intersected to find the number of patents common to both. Out of a possible 3,250 (the size of the smaller set) only 60 "common" patents were found to belong to both sets. That is, about 5 per cent of the 3,250 citing patents were themselves cited by one or more of the 3,250 patents.

"Consistency" of Classification

Classifications of citing patents were compared with classifications of the cited patents and about 46 per cent of the citing patents were in a different class from the corresponding cited patents.[3]

## The Experimental System and Its Application to the Citation Index.

### Description

This patent citation index has been prepared for use in the AMNIP system described in Ref. (5). The AMNIP system is a general purpose man-machine system specified by M. Kochen and programmed by F. Blair and D. Reich. The memory for this system is conceived as a graph which, in the citation index application, would be a bi-directional graph in which the "nodes" are interpreted as patent disclosure numbers, and the "links," are "cites," and "is cited by," indicating the direction of search as forwards or backwards in time. Classification codes are also viewed as nodes, with the relevant link "is classified under." Programs for this system permit the user, seated at a keyboard console, to form a complex query consisting of conjunctions, disjunctions, and negations of nodes, specify the kind of link desired, and obtain a list or display of patent numbers and/or classes that satisfy the query.

### Use

### Complex Querying

This system could be used to permit more complex queries than are possible in the usual bound-book citation index. For example, an attorney could ask for all patents that cite patent a and patent b but not patent c, or for patents that are cited by patent d. He could further specify the class of the desired patents, e.g. dis-

play all patents that cite patent a except those in class g.   The
ability to perform such complex searches can be expected to speed
up the lawyer's search.   (He could avoid inspecting patents known
a priori to be of no use to his search.   He could also avoid inspecting some  patents more than once--if they occur in the citation
listing of two or more patents in the search query.)

Path Tracing

Another useful facility of this system would be the ability to
trace paths through the network.   Iterative combinations of the basic system programs could be used to find 2nd, 3rd, or nth generation ancestors and descendants (find all patents connected to a, then
those connected to the patents just found, etc.).   A variety of
"routes" could be traced through the citation net.   One could, for
example, trace both forwards in time to citing patents, backwards
in time to cited patents, or some combination of both.   This path
tracing could be either completely automatic or under user control,
with the attorney eliminating undesirable paths at each iteration.
With automatic path tracing, we would intuitively expect an increase
in the number of relevant documents obtained at a cost of more
trash.   However, the trade-off between path-length (number of gen-
erations) and retrieval effectiveness (hit-trash rates) is as yet unknown.   If it is found that automatic tracing is too costly (in terms
of trash), then selective tracing in the man-machine system could
be substituted, with the attorney eliminating undesirable paths at
each step.

Weighting

Other techniques for controlling trash in a machine system are
conceivable.   One could introduce a weighting scheme, ranking patents, for example, on the basis of the number of citations to the
patent or the number of claims in it.   The highest ranking patents
would then be traced first.   The assumption, however, underlying
such an approach is that there exist "key" patents which are more
likely than others to be of value, and that such keyness is system-
dependent--dependent only on the relation of one patent to another

in the system--rather than query-dependent--relative principally to
the specific search in question.

It must be cautioned, however, that the extent to which any of
the above facilities (complex querying, path tracing, weighted tracing) would actually be used in a system must remain a matter of
speculation until such a system is built.   Predicting the utility of a
tool before the availability of the tool--particularly when a training
period is required--is guess-work at best.

It should also be cautioned that this kind of machine stored
citation index--only patent numbers and classification codes stored
in memory--should be supplemented with a patent display device.
Use of the citation index alone, without provision for examining the
patents easily and quickly, could be quite irritating.   The attorney
would have to take his list of patents to be examined and physically
obtain each patent or patent abstract.   Each patent would probably
be stored in a different location, many would not be immediately
available, and quite possibly, after inspection, most would be of
little interest to him.

By storing the patent, or the abstract, in a display device for
immediate inspection, the inconveniences of physical acquisition at
any rate could be greatly reduced.   According to expert opinion,
such display devices are technologically and probably economically
feasible.

Experimentation

While awaiting the availability of the system for large scale
testing, a pilot test of the citation index was performed in collaboration with IBM patent attorneys at the Poughkeepsie IBM plant.

For this experiment, a list of 60 "search" patents, together
with their references  (cited patents), were supplied by the attorneys
for tracing through the citation net.   These patents constitute the
question set.   Two "routes" through the network were traced for
the experiment.   (1) Simple Citation Route--in which forward-in-
time patents immediately connected to the search patent were obtained, and (2) Reference-Citation Route--in which references to the

search patent were found by tracing backward-in-time and then these references traced forward to obtain second generation patents. The patents found in these tracings constitute the answer set.

　　　The patents obtained by tracing through the net were returned for evaluation to the individual attorneys, who were requested to evaluate them on a four-point scale. Each attorney judged the answer set (patents found by tracing through the net) for his own questions.

　　　A control group of patents was included in the lists of patents submitted for evaluation. These control patents for each question were selected from the same class as the question (search) patent. IBM used a classification system with major and minor subdivisions. The controls were chosen from the minor class--the narrowest subdivision in which the search patent fell.

Trace Results

　　　In all, of the 60 search patents submitted, 43 patents, together with their associated traces, were found. This was a surprisingly large number since the data were fairly old, certainly not comprehensive, and no known constraints were placed on the type of patent to be submitted for searching. Forty of the 43 were evaluated and returned by the attorneys. These 43 patents were then our "questions." For the simple citation route (patents forward in time from the search patent) there were 11 questions, and 32 for the second-level reference-citation search (back to the references and forward from these). The total number of "answers" (traced patents) found were respectively, 91 for the simple citation route and 735 for the second level search. All patents found in the first level search were sent to the attorneys for evaluation and a sample of 240 were sent for the second level search. The control group was approximately 10-15 per cent of the size of the experimental group. These data are tabulated in Table 1.

　　　The selection of questions and associated answers was clearly not under control of any sort. Whatever questions corresponded to data in our citation index were chosen for the experiment. It was clearly not possible to vary the classes in which the patents fell, the age of the patents, the number of claims in the patent, or the evaluators. As it turned out, all the first level searches were questions from the same attorney. Thus, although we had 91 observations, only one attorney evaluated these patents. While this does eliminate the problem of variability between different evaluators, it also makes generalization untrustworthy. The results obtained from the first-level search must consequently be understood as the results of a pilot test only.

| | Number of "questions" sent | Number of "answers" found | Number of "answers" sent | Number of controls sent |
|---|---|---|---|---|
| First level search | 11 | 91 | 91 | 15 |
| Second level search | 32* | 735 | 240 | 29 |

*see Note 4

Table  I

## Evaluation

Evaluation of a citation index can be viewed in two ways. We can focus either on the "questions" or on the "answers." In the first, the test population is the set of questions, and the variable to be measured, the proportion of relevant documents per question. In the second, the answers are the population and the variable, the proportion of relevant documents in the population--with differences between questions ignored. For the first, one would determine the proportion of relevant documents per question and then average the proportions; for the second, one would, essentially, weight the individual proportion per question by the number of items in the answer set. The choice between the two approaches is arbitrary. For this experiment, the second view was taken.

We thus want to ask the following questions:

(1)   Does the proportion of relevant answers in the citation population exceed the proportion in the control (classification) population? If so, how much?

The answer to this question should indicate--given a choice between searching in a citation index or in this narrow classification system (the controls were patents chosen from the same class as the search patent) which should one choose.

And a second, supplementary, question should also be asked:
(2)    What proportion of patents judged to be relevant would not have been found by searching in the same class as the question patent?

The answer to this question should indicate the extent to which a citation index could supplement (rather than substitute for) a conventional search in this classification system.

## Relevance

Few tests of information retrieval tools, lately, can avoid discussing the problem of "relevance." The difficulties, as usually stated, are: (1) there is little agreement between people in the judgment of relevance, (2) the relevance of a document to a query admits of degrees, and (3) the notion of relevance itself is vague and "the question of the meaning of relevance is, in many ways, obscure."

The first problem--agreement between different people in the judgment of relevance--we avoid by letting the person who asks the question evaluate the answer. The assumption here is that relevance is a subjective judgment, and we can do no better than evaluation by the "subject" in the best position to judge.

The second problem--the degrees of relevance of a document to a query--we attempt to detect using the four-point relevance scale (no relevance, slight, some, great relevance).

The third problem--the vagueness of the notion of relevance--is both a real problem and a problem of our own creation. In this experiment, the problem is mitigated because we are running a comparative test. In whatever manner the notion of relevance is interpreted by each evaluator, we can at least assume him to be consistent with himself. His interpretation of the notion of relevance for evaluating the relevance of a citation patent to his question should not differ from his interpretation for the control patent. (Of course, the evaluator was not told which patents were controls.)

In general, however, this problem of the vagueness of the relevance notion is to some extent an artificially created one.    The designer of an IR test has instructed the experimental subjects to tell him "Is this document relevant to this question?"   He then claims that the notion of relevance, (i. e. the instructions he has given) is obscure!   One wonders, then, why the instructions were given in this manner.   By reformulating the instructions to the evaluators we might presumably make these instructions less vague and obtain more reliable--and perhaps less variable answers.   To do this, we would ask evaluators to indicate, for example, whether (in an infringement-type search):

(1)     This patent alone contains all information desired (i. e. clearly indicates infringement).

(2)     This patent contains some information (i. e. this patent, together with others, clearly indicates infringement).

(3)     This patent alone contains information from which infringement could be inferred or argued. . . . etc.

Different sets of such differentiated answers might be constructed for different types of searches (i. e. novelty, infringement, etc. ).

In other words, by giving our users more precise directions, we may get more precise answers.

Results

For the first level search (ordinary citation index search), the number of patents judged relevant are as indicated in Table II below.

|  | No relevance | Slight relevance | Some relevance | Great relevance | Total |
|---|---|---|---|---|---|
| Citation search | 58 | 30 | 3 | - | 91 |
| Classification search | 12 | 3 | - | - | 15 |

Table II

Since the results were, in general, low, we combined the "slight, " "some" and "great" categories to obtain, Table III.

| | Number not relevant | Number relevant | Total No. of observations | Proportion relevant |
|---|---|---|---|---|
| Citation | 58 | 33 | 91 | $\frac{33}{91}$ = . 36 |
| Classification | 12 | 3 | 15 | $\frac{3}{15}$ = . 20 |

Table III

We tested, then, the significance of the difference between the two proportions according to:

$$Z = \frac{p_1 - p_2}{\sqrt{\dfrac{n_1 \sigma_1 + n_2 \sigma_2}{n_1 - n_2 - 2}}}$$

where $p_1$ = proportion of citations relevant = . 36

$p_2$ = proportion of controls relevant = . 20

$n_1$ = number of citation observations = 91

$n_2$ = number of control observations = 15

Results were significant at the 99 per cent level.

We wish to reiterate our earlier caution that these results are to be interpreted as a pre-test, only.

We conclude then, that, within the limitations of this test, it is indicated that searching in a citation index would yield a significantly higher proportion of relevant patents than searching, in this classification system, within the same class as the "question" patent.

The secondary question we now ask is, how much could we supplement a search in the classification system with a search in the citation index? To answer this, we ask--of those patents judged to be relevant from the citation index search--what proportion were not in the same class as the search patent? These data are indicated in Table IV, below.

Thus, roughly 50 per cent of the patents judged to be relevant would not have been found by searching in the classification system (in the minor class) alone.  No significance test was run on this question for the following reason.  We would have to assume that, if an observer had been presented with only those patents in the smaller set of patents from the different classes, his judgment of relevance would have been the same as in this test--in other words, that the relevance judgment is not dependent on the number of items to be judged.

| Number of relevant patents | Number in a different class from the search patent | Proportion |
|:---:|:---:|:---:|
| 33 | 16 | $\frac{16}{33} = .48$ |

Table IV

In view of the already limited nature of this experiment, imposing an assumption of this kind seemed unwarranted.  We thus present the data as indicative only without testing significance.

All the above data refer to the first level search.  Results of the second level search were obtained, 240 observations in all, but the results were not significant.

Acknowledgment

I wish to acknowledge the help of M. Kochen for initiating this work and for his general encouragement, of C. T. Abraham for his very patient advice in all matters statistical, of A. H. Dickenson for preparing the data for the citation index and for classifying many of the patents for this study, and of the patent attorneys who acted as evaluators of the patents found.

Notes

1.  This work was supported in part by the United States Air Force under Contracts AF19(626)-10, AF19(628)-2752 and AF30(602)-3303.  Portions of this paper appeared in Automation and Scientific

Communication, Proceedings of the ADI Annual Meeting, 1963, under the title "A Machine-Stored Citation Index to Patent Literature--Experimentation and Planning."

2.    Prepared by A. H. Dickinson of the IBM Patent Department who kindly made them available to us.

3.    These data are based on an extremely small sample, the 60 patents in the intersection of the set of cited and citing patents. These were the only patents for which classification of both citing and cited patents were available.

4.    29 returned by evaluators.

## References

(1)    Atherton, P. and Yovich, J. C., "Three Experiments with Citation Indexing and Bibliographic Coupling of Physics Literature," American Institute of Physics, April, 1962.

(2)    Garfield, E., "Citation Indexes for Science," Science, July 15, 1955, pp. 108-111.

(3)    Garfield, E., "Breaking the Subject Index Barrier--A Citation Index for Chemical Patents," Journal of the Patent Office Society, August, 1957.

(4)    Kessler, M. M., "Technical Information Flow Patterns," Proceedings of the Western Joint Computer Conference, 1961, pp. 247-257.

(5)    Kochen, M., Abraham, C. T., and Wong, E., "Adaptive Man-Machine Concept Processing," IBM Final Report on Contract AF19(604)-8446 for Electronics Research Directorate, Air Force Cambridge Research Laboratories, June, 1962.

(6)    Little, Arthur D., Inc. "Centralization and Documentation," Final Report to the National Science Foundation, C-64469, July, 1963.

(7)    Lipetz, Ben-Ami, "Compilation of an Experimental Citation Index from Scientific Literature," American Documentation, 1962, pp. 251-266.

(8)    Tukey, J. W., "The Citation Index and the Information Problem, Opportunities and Work in Progress," Annual Report for

1962 under National Science Foundation Grant NSF G-22108, Statistical Techniques Research Group, Princeton University.

III C.   Toward Controlled Experiments in the Construction of an
Adaptive Man-Machine Associative Memory for Information Retrieval

C.  T.  Abraham

We shall propose an abstract model for an associative[1] memory, specify in detail an experimental design for the evaluation of
such a model for a specific data base.   The data base to be used
is the one prepared for the AMNIP system.

The theoretical model is one which uses, with advantage, the
man-machine system, since it exploits not only the relations that
exist between the name-predicate-name sentences[2] on the basis of
common R's and L's but also the patterns of relations established
by humans in the actual usage of the system for fact retrieval.
(An acquaintance with the AMNIP system proposal in "Adaptive Man-
Machine Concept Processing" is helpful.)

The theoretical model we propose is based on a percolation
process and a brief description of this kind of stochastic process
will be given.   For a more detailed description see References (1)
and (3).   A percolation process, as the name implies, describes
the flow of a liquid through a crystal medium.   In this context, a
crystal is defined as a set of atoms together with the bonds between
the atoms.   A bond is a directed link between two atoms.   The random (probabilistic) aspect of this process is introduced by imposing
a probability q on each of the bonds so that at any instant there is
a probability q that a bond is closed or dammed against the flow of
a liquid introduced at an atom, say, A, of a directed bond from
atom A to any other atom B.   There is no restriction on the number
of bonds (as long as this number is finite) which can exist between

174

two atoms.   Also, for any two atoms A and B, it is possible to
have bonds from A to B as well as from B to A.   For example, we
can define a crystal whose atoms are located at the integer points
of a doubly infinite line.   We could have each even numbered atom
connected by a bond to its immediate neighbors on the left and right
and each odd numbered atom by a one-way bond to its neighbor on
the right.   See Figure 1 below.

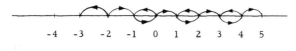

-4   -3   -2   -1   0   1   2   3   4   5

Fig. 1

A directed path from any atom A to another atom is called a
walk from A.   A walk in which the path does not return to any one
of the atoms is called a self-avoiding walk or a nonredundant path.
We shall indicate later an algorithm based on a special kind of partition of natural numbers which will greatly facilitate the tracing of
nonredundant paths.   At that time, we shall also give certain formu-
lae which will give the number of redundant and nonredundant paths
in a crystal.   An n-step self-avoiding walk from an atom A will be
denoted by $S_n(A)$ and $f_n(A)$ will denote the frequency of n-step self-
avoiding walks from an atom A.   All atoms which have the same
number of n-step self-avoiding walks, for n = 1, 2, 3, . . . , are
said to belong to an out-like class.   There are three axioms to be
satisfied by any crystal in order that percolation theory is applicable
to the crystal.   These axioms are:

1.     Each atom should have only a finite number of bonds.
2.     Each atom should belong to exactly one of a finite number of
       out-like classes, say, $C_1$, $C_2$, . . . $C_k$.
3. a   In any finite subset of atoms, there is at least one atom from
       which there is a bond to an atom outside the set.
3. b   If a subset of atoms does not contain atoms of at least one
       out-like class, then this subset contains an atom from which
       a bond leads to some atom not in the subset.

If $f_i(n)$ is the frequency of distinct n-step self-avoiding walks of an atom in the out-like class $C_i$, and if

$$f(n) = \max_{1 \le i \le k} f_i(n) \tag{1}$$

where k is the total number of out-like classes and if

$$\lambda = \lim_{n \to \infty} \frac{1}{n} \log f(n) \tag{2}$$

then, $\lambda$ is called the critical constant of the crystal. In addition

$$\lambda = \lim_{n \to \infty} \frac{1}{n} \log f(n; r(n)) \tag{3}$$

if     $\lim_{n \to \infty} \dfrac{r(n)}{n} = 0$

where r(n) is the number of self-avoiding walks in an n-step walk.

As indicated earlier, the random aspect of the percolation process is introduced by assigning a probability q of damming a bond. Obviously $p = 1-q$ is the probability that a bond is open. The main theorem in any percolation process concerns the probability with which a liquid introduced at an atom A as the source atom will flow to other atoms. More specifically, this theorem gives a critical value $p_c$ of p such that when $p \le p_c$, a liquid introduced at a source atom A will flow to only a finite number of atoms. Thus if $C(A, p)$ denotes the probability for the liquid introduced at atom A to reach only a finite number of atoms, then the theorem can be stated as follows:

Theorem  If $p \le p_c$, then $C(A, p) = 1$ \hfill (4)

The critical probability $p_c$ is related to the critical constant $\lambda$, as we shall see in the sequel. Thus for a given crystal, it is possible to obtain bounds for the critical probability. It can be shown that if A and B are any interconnected atoms of a crystal satisfying the three axioms, then:

$$C(A, p) = C(B, p) \tag{5}$$

Thus for any interconnected crystal, $C(A, p)$ is the same for every atom A.

Permeability is the twin problem to percolation.  Here the liquid is introduced at certain atoms called "surface atoms" and we are interested in the problem of the liquid "wetting" an atom in the "interior" of the crystal.  To describe permeability, definitions are necessary.  Thus the n-set of an atom A are all those atoms from which the shortest self-avoiding path to A consists of n-steps.  If $W_n(A, p)$ denotes the probability that a liquid introduced into the n-set atoms of A will wet A when p is the probability for a bond to be open, then if A and B are any two atoms such that $W_n(A, p) = W_n(B, p)$, then A and B are said to belong to the same in-like class. Thus an in-like class of atoms is a set of atoms in which every pair of atoms is in-like.  An atom A is said to be an enclosed atom of a set C if there exists an integer n such that every $S_n(A)$ passes through at least one atom of the class C.  The interior of a class C of atoms is the set of all enclosed atoms of C which do not belong to C.  A set of atoms with non-empty interior is called a boundary of that interior.  An atom A is n steps away from a set C of atoms if some $S_n(A)$ but no $S_{n-1}(A)$ pass through an atom of C.  A crystal having the property that when the direction of the bonds is reversed, the resulting crystal also satisfies the three axioms required of a crystal, is called a reversible crystal.  A reversible maze is a maze obtained by reversing the direction of all bonds in the original maze, leaving dammed bonds dammed.  Atoms in a reverse maze or crystal are denoted by the same letters as in the original one, with a prime.  Thus A' in the reverse maze or crystal corresponds to A of the original maze or crystal.  Let $W_n(A, p)$ denote the probability that atom A is wet when every atom in its n-set is a source atom.  (The only source atoms are the members of the n-set.)  $W_n(A, p)$ is monotone decreasing in n and bounded by one.  Therefore, $\lim_{n \to \infty} W_n(A, p) = W(A, p)$ exists.  We shall give below a theorem without proof relating percolation and permeability.

<u>Theorem</u>   $W(A, p) + C(A', p) = 1$                                 (6)

for any maze and its reverse maze.

Theorem   If $\mu = \underset{A}{\text{Sup}} \; \underset{n \geq 1}{\text{inf.}} \; [\; \frac{1}{n} f_A(n) \;]$, then the critical

probability $p_c \geqq e^{-\mu}$                                         (7)

In some crystals, $\lambda = \mu$. This happens specifically in crystals
where

$$f_i(n) = e^{\lambda n + 0(n)} \quad \text{where} \quad \frac{0(n)}{n} \to 0 \tag{8}$$

Thus if $p \leq e^{-\mu}$, then $C(A, p) \leq 1$. If $p > p_c$, then $C(A, p)$ is
obtained as a solution of an equation of the form $x = F(x)$ involving
certain generating functions.

In the light of the above brief description of percolation processes, the definitions and problems of the construction of an associative memory for the AMNIP system can be briefly stated as follows:

I        The atoms of the crystal are the names and the bonds between the names are predicates. The rules by which bonds are established would depend on the type of information retrieval   required
of the system. Thus we might look at various crystal structures
each of which may have only one or a subset of the set of the predicates as bonds. Thus we may have a crystal in which the atoms
are names of authors and the bond is the predicate "coauthored
with." Again, citations form another crystal. Both "coauthored
with" and "cites" could be jointly used as bonds in another crystal.
Which predicate or predicates form bonds will have to be decided
either on the basis of actual usage requirements or could be arbitrarily decided in advance. Thus with the data in RLR form, we
can form various crystal structures.

II       The crystal structures formed as in I will have to be examined to see whether they satisfy the three axioms required of a
crystal for percolation process. The three axioms reduce to checking the following:

a)   Does each name have only a finite number of bonds?

b) Does each name belong to exactly one of a finite number of out-like classes?

c) Does any subset of names include a name from which there is a bond to a name outside this subset?

For the RLR data, these axioms will be definitely satisfied for at least some bonds. If for any crystal structure, any one of these three axioms is violated, then the criteria for establishing bonds have to be altered or certain names or predicates or both have to be removed and these considered separately.

III     All the interconnected names have to be listed.

IV     All the names have to be classified into their out-like classes.

V     When new names are added, they have to be classified as in III and IV.

VI     When humans use the system, they enter the system by providing a name or a list of names and then select a subset of the set of names which are linked to the name or names with which they enter the system. There is a selection by a process of elimination on the part of humans to find an answer to the query. This is equivalent to saying that the manner of the use by humans imposes a probability with which bonds are used or not used. Thus for different queries, there are bound to be different probabilities with which the bonds leading from the source names are used. The probability with which any bond will be used in a particular query can be estimated by the ratio of the number of links that are actually used to the total number of links that lead from the source name or names. For various queries, these probabilities will be estimated for the various source names. Thus for each name, there is a sequence of probabilities that its bonds will be open. If for each name, the maximum of these probabilities is computed and if the supremum of all these maximum values, say, $p$ is obtained, then as long as $p$ is less than the critical probability of the crystal, we are assured that it is certain that for any query the number of links to be traced is finite. Thus the critical parameter of the

system is this p.   It is quite likely that p may exceed the critical
probability.   If this is the case, then there are two possibilities
open.   They are:

   1.   Separate those source names for which the maximum probability exceeds the critical value from the rest of the names
and have special identifications for them.   This is equivalent
to saying that whenever these names are used as source names,
the system user should be told that search for links in great
depth has to be expected.   Thus, unless quite necessary, the
human can be discouraged from using these names as source
names.

   2.   For all the names, a value of p can be found such that
with a probability almost one, $(1 - \epsilon)$, p will be below the
critical probability.   In this approach, we allow the user to
use any source name and we do not indicate the depth of
searching involved but achieve reasonable search time and retrieval in the long run.

VII     The problem of permeability is quite relevant to the associative memory in the following sense.   Suppose a human enters the
system with a set of names and wants to trace all the connections
these names have.   Before permitting such a request, a check has
to be made as to whether these names acting as the atoms in the
surface, will allow penetration to a great depth.   This question can
be answered very simply by considering the reverse crystal.
[ If $C(A', p) = 1$, then $W(A, p) = 0$ ]

   It is hoped that as a result of these investigations there will
emerge an organization of the memory into several levels.   There
is one level where all the names are stored as simple RLR sentences.   Then there is another level at which sentences are stored
in $(R_i L_j R_j L_l R_m \ldots)$.   It is also hoped that these strings of R's
and L's will themselves act as nodes (concepts) for another graph
and in such a graph, it will be possible to form organizations at
two levels as before.

Appendix

## Algorithms for tracing nonredundant paths

Ross and Harary [Ref. (4)] have given some techniques using matrix operations for determining the number of redundant and non-redundant paths of various lengths in a directed graph. We shall briefly describe these techniques which are very suitable for small directed graphs. Then we shall indicate how the relations which exist among redundant paths can be used to provide heuristics for efficient path tracing.

If N is the nxn connection matrix of a directed graph with n vertices or nodes, the number of nonredundant paths of length $r \leq n$ that exist between the n nodes is given by

$$P_r = (N^r - R_r) - d(N^r - R_r) \tag{9}$$

where $P_r$ is an nxn matrix whose $(i,j)^{th}$ element is the number of nonredundant paths of length r between the nodes i and j $(1 \leq i \neq j \leq n)$ and d(M) is the diagonal operator of any matrix M; i.e. a matrix whose diagonal entries are those of the matrix M and whose nondiagonal entries are all zeroes. $R_r$ is the matrix whose $(i,j)^{th}$ element is the number of redundant paths of length $r \leq n$ between the nodes i and j $(1 \leq i,j \leq n)$.

The matrix $R_r$ can be determined by considering certain restricted partitions of the natural number r. Involved in the formula for $R_r$ is the adjacency matrix NxN', where N' is the transpose of N and the multiplication is elementwise. For a more detailed discussion of the derivation of $R_r$, the reader is referred to the paper by Ross and Harary [Ref. (4)].

The redundant paths in a directed graph are related to one another and the occurrence of certain redundant paths will preclude the occurrence of certain other paths. In fact, the information about such inconsistent paths can be used to provide heuristics for efficient path tracing algorithms. We shall now discuss the procedure for determining inconsistent redundant paths. Redundant paths of any specified length r correspond to restricted partitions of the natural number r.   These partitions are the number of ways in which r can be expressed as the sum of three non-negative integers so that the first and the third integers are not both zero and the second integer is greater than one and the order of the summands is significant.   If $r = r_1+r_2+r_3$, then a redundant path from any vertex i to any other vertex j is said to satisfy this partition of the number of edges from the initial vertex i to the place where any repeated vertex occurs for the first time is $r_1$, the number of edges from the first appearance of a vertex to its repeated appearance is $r_2$ and the number of edges from this repeated appearance to the terminal vertex j is $r_3$.   Thus the redundant path i j i j corresponds to the partition 0+2+1 of the number 3 when regarded as a path from i to j.   The maximum number of restricted partitions of r is

$$2 \begin{pmatrix} [\frac{r+1}{2}] \\ 2 \end{pmatrix}$$

and the exact number of restricted partitions of the type described above is    $\binom{r}{2} - 1$    .   There are both consistent and inconsistent partitions.   Two partitions are consistent if paths satisfying both partitions are possible.   Thus if we consider two paths of length 4 from i to j, and if $p_1$ and $p_2$ are partitions corresponding to the paths i j k l j and i k l i j respectively, then the most general path satisfying $p_1$ and $p_2$ is i j k i j and such a path may exist.

| Partition | Path |
|---|---|
| $p_1$: 1+3+0 | l j k m j |
| $p_2$: 0+3+1 | i l m i n |
| path redundant in j<br>path redundant in i ] | i j k i j |

Examples of inconsistent paths

| Partition | Path |
|---|---|
| $p_1'$:   2+2+0 | i k j l j |
| $p_2'$:   1+3+0 | i j k l j |

The general path satisfying both $p_1'$ and $p_2'$ is i j j l j, but this path is not possible.

Thus in tracing paths of various lengths between i and j, the occurrence of certain redundant paths will preclude the occurrence of certain other redundant paths. Thus for redundant paths of any specified length r, it is possible to construct a table of consistent and inconsistent partitions. This notion of inconsistency can be extended to more than a pair of partitions. We shall give below tables of consistency for paths of lengths 4, 5 and 6.

### Table of consistency for paths of length 4

|       | $p_2$ | $p_3$ | $p_4$ | $p_5$ |
|-------|-------|-------|-------|-------|
| $p_1$ | 1     | 0     | 1     | 0     |
| $p_2$ |       | 1     | 0     | 0     |
| $p_3$ |       |       | 0     | 1     |
| $p_4$ |       |       |       | 1     |

Where the $(i, j)^{th}$ entry is one if partitions $p_i$ and $p_j$ are consistent, the partitions in the table are as follows:

| Partition        | Path      |
|------------------|-----------|
| $p_1$ = 1+3+0    | i j k l j |
| $p_2$ = 0+3+1    | i k l i j |
| $p_3$ = 2+2+0    | i k j l j |
| $p_4$ = 0+2+2    | i l i k j |
| $p_5$ = 1+2+1    | i k l k j |

From the above matrix, the following table of partitions which are inconsistent when taken three at a time is possible.

| Partitions     | Consistency |
|----------------|-------------|
| $p_1 p_2 p_3$  | 0           |
| $p_1 p_2 p_4$  | 1           |
| $p_1 p_2 p_5$  | 0           |
| $p_2 p_3 p_4$  | 0           |
| $p_2 p_3 p_5$  | 0           |
| $p_3 p_4 p_5$  | 0           |

Thus if i j k l j and i k l i j are both redundant paths between i and j, then the only other redundant path between i and j of length 4 is i l i k j.

The above illustrative example indicates how the restricted partitions for redundant paths of any desired length can be formed and how by constructing consistency tables, it is possible to reduce the amount of work involved in path tracing.

The partitions and consistency tables for redundant paths of lengths 5 and 6 are listed below.

### Redundant paths of length 5

| Path | Partition |
|------|-----------|
| iklmj | $p_1 = 1 + 4 + 0$ |
| iklmij | $p_2 = 0 + 4 + 1$ |
| iljmkj | $p_3 = 2 + 3 + 0$ |
| iklimj | $p_4 = 0 + 3 + 2$ |
| ikljmj | $p_5 = 3 + 2 + 0$ |
| ililmj | $p_6 = 0 + 2 + 3$ |
| ijkjlj | $p_7 = 1 + 2 + 2$ |
| ilkmkj | $p_8 = 2 + 2 + 1$ |
| iklmkj | $p_9 = 1 + 3 + 1$ |

### Pairwise consistency matrix for paths of length 5

|       | $p_2$ | $p_3$ | $p_4$ | $p_5$ | $p_6$ | $p_7$ | $p_8$ | $p_9$ |
|-------|-------|-------|-------|-------|-------|-------|-------|-------|
| $p_1$ | 1 | 0 | 1 | 1 | 1 | 1 | 1 | 0 |
| $p_2$ |   | 1 | 0 | 1 | 1 | 1 | 0 | 0 |
| $p_3$ |   |   | 1 | 0 | 0 | 0 | 0 | 1 |
| $p_4$ |   |   |   | 0 | 1 | 0 | 1 | 1 |
| $p_5$ |   |   |   |   | 1 | 1 | 1 | 1 |
| $p_6$ |   |   |   |   |   | 1 | 1 | 1 |
| $p_7$ |   |   |   |   |   |   | 1 | 1 |
| $p_8$ |   |   |   |   |   |   |   | 0 |

The partitions for redundant paths of any finite length can be obtained and pairwise consistency tables can be made.

Thus for tracing all paths of a given length from node i to node j, it is possible to employ the consistency tables to supply a program heuristic to reduce the machine time for path tracing.

### Redundant paths of length 6

The total number of partitions = $[ \binom{6}{2} - 1 ] = 14$

| Partitions | Paths |
|---|---|
| $p_1 = 0 + 5 + 1$ | iklmsij |
| $p_2 = 0 + 4 + 2$ | iklmisj |
| $p_3 = 0 + 3 + 3$ | iklimsj |
| $p_4 = 0 + 2 + 4$ | ikilmsj |
| $p_5 = 1 + 5 + 0$ | ijklmsj |
| $p_6 = 1 + 4 + 1$ | iklmskj |
| $p_7 = 1 + 3 + 2$ | ijkljmj |
| $p_8 = 1 + 2 + 3$ | ijkjlmj |
| $p_9 = 2 + 4 + 0$ | ikjlmsj |
| $p_{10} = 2 + 3 + 1$ | ilkmskj |
| $p_{11} = 2 + 2 + 2$ | ikjljmj |
| $p_{12} = 3 + 3 + 0$ | ikljmsj |
| $p_{13} = 3 + 2 + 1$ | iklmsmj |
| $p_{14} = 4 + 2 + 0$ | iklmjsj |

The consistency matrix for redundant paths of length 6

| | $p_2$ | $p_3$ | $p_4$ | $p_5$ | $p_6$ | $p_7$ | $p_8$ | $p_9$ | $p_{10}$ | $p_{11}$ | $p_{12}$ | $p_{13}$ | $p_{14}$ |
|---|---|---|---|---|---|---|---|---|---|---|---|---|---|
| $p_1$ | 0 | 1 | 1 | 1 | 0 | 1 | 1 | 1 | 0 | 1 | 1 | 1 | 1 |
| $p_2$ | | 0 | 1 | 1 | 1 | 0 | 1 | 1 | 0 | 1 | 1 | 1 | 0 |
| $p_3$ | | | 1 | 1 | 1 | 0 | 1 | 1 | 0 | 0 | 1 | 1 | 0 |
| $p_4$ | | | | 1 | 1 | 1 | 1 | 1 | 0 | 0 | 0 | 1 | 1 | 1 |
| $p_5$ | | | | | 0 | 1 | 1 | 0 | 1 | 0 | 1 | 1 | 0 |
| $p_6$ | | | | | | 1 | 0 | 0 | 1 | 0 | 1 | 1 | 1 |
| $p_7$ | | | | | | | 0 | 0 | 1 | 0 | 1 | 1 | 1 |
| $p_8$ | | | | | | | | 0 | 1 | 0 | 1 | 0 | 0 |
| $p_9$ | | | | | | | | | 0 | 1 | 0 | 1 | 1 |
| $p_{10}$ | | | | | | | | | | 0 | 1 | 1 | 1 |
| $p_{11}$ | | | | | | | | | | | 0 | 1 | 1 |
| $p_{12}$ | | | | | | | | | | | | 0 | 0 |
| $p_{13}$ | | | | | | | | | | | | | 1 |

## Notes

1.    See also paper III F of this volume.
2.    See paper I A for more details.

## References

(1)   S. R. Broadbent and J. M. Hammersley. "Percolation Processes, Crystals and Mazes." Proc. Camb. Philosophical Soc. Vol. 53 (1957), pp. 629-641.

(2)   J. M. Hammersley. "Percolation Processes, the Connective Constant." Vol. 9 (1961), pp. 533-543.

(3)   T. E. Harris. "A lower bound for the critical probability in a certain Percolation Process." Proc. Camb. Philosophical Soc. Vol. 56 (1960), pp. 13-20.

(4)   L. C. Ross and F. Harary. "On the determination of redundancies in Sociometric Chains." Psychometrika. Vol. 17, No. 2, 1952.

## III D.  A Proposed System for Multiple Descriptor Data Retrieval[1]

### W. D. Frazer

I.  Introduction

A problem encountered frequently in information retrieval may be described as follows:

Consider a file composed of items which are t-tuples of symbols: $(A_1, A_2, \ldots A_t)$ where each $A_i$ can assume only one of $k_i$ distinct values $a_{i1}, a_{i2}, \ldots a_{ik_i}$. It is desired that, having specified a value for each of some subset of the "fields," $A_i$, one be able to recover from the file all items whose field values for the chosen subset of fields coincide with the values prescribed.

$A_1$ might correspond, for example, to a date, $A_2$ to a location, etc. Then $k_1$, the number of distinct values possible for $A_1$, would probably differ from $k_2$, the number of distinct values possible for $A_2$, and also from $k_3$, $k_4$, . . . . In virtually all data processing applications, each field $A_i$ is represented by a string of bits; different strings, all of the same length, correspond to the various $a_{ij}$, as j varies from 1 to $k_i$.

There have been suggested several different approaches to this problem--some involving the use of unusual devices, and some relying on unusual storage techniques. A brief summary of these approaches follows:

The most efficient method of locating the items in a conventional store so as to be able to extract all desired items with a minimum of search effort is to have each item filed as many times as there are data fields (t in this case). These files may be called topic files, each organized on the basis of a different data field.

One then examines the incoming "query" and determines that field
specified by the query whose corresponding topic file contains the
fewest items.    One then searches this topic file on an item-by-item
basis.    This is the card catalog approach; the storage requirements
for such a system are clearly very large.

An alternative approach, one requiring somewhat less storage
space, is to store the topic file in the form of tables; one would have a
table corresponding to the value $a_{ij}$ for field $A_i$, for example, which
would list the location of any item having that value for field $A_i$.    Retrieval of items with the desired characteristics would thus require the
comparison of all such lists corresponding to the prescribed field values.
As before, it would be desirable to check the list with the fewest entries
against the others.    This is the "coordinate indexing" approach. [1]

A third approach involves the use of "associative" or "content-addressable" memories.    Here the data file is searched in its entirety virtually simultaneously, and items which satisfy the given conditions are
"tagged" in some fashion.    The tags are then searched sequentially and
"tagged" items removed. [3]    The main disadvantages attributed to
this method seem to be connected with the amount of equipment required
to "tag" the items and the necessity for a sequential search (alternatively
more equipment) to remove tagged items.

The final approach is simply to store each t-tuple, ordering the
file on the basis of the most frequently queried field if possible, but
at least on some likely field, and to search the file exhaustively for
queries not specifying the "ordering field."    For queries specifying this
"ordering field," a binary search is first performed to narrow the candidate items to those whose "ordering field" values correspond to that
of the query; this set of items is then searched exhaustively.

It is the purpose here to investigate some possibility of alternative
solutions to this problem.    All of the above solutions entail the construction of some kind of "image" of the data in the file.    In the first two cases,
the image is formed at the time the data enters the file.    In the third
case, the image is variable and is constructed by hardware to meet the
specification of the query--the image is the set of tags.    In the last case,
the image is the trivial one consisting of the data itself.

II.  Searching and Hardware Images

Only in the case of the content-addressable memory above is any attempt made to employ hardware to form the image of the data. The idea of hardware-formed images seems to offer great potential for rapid search, and so bears further investigation. Rather than require that the hardware-formed image be the answer to the query, however, we shall investigate the possibility that some combination of search techniques and hardware-formed images might provide a solution. That is, one might form an image in such a way that a search on the image would give information as to when to look in the main store for desired items. One would then search the main store and retrieve the desired items. In order to succeed, such a system should insure that the total number of searches--image plus items in file--be less than the number of items in the file.

As a first attempt, consider the possibility of forming an image by forming some logical function of adjacent data words in the store. It is reasonable to use address adjacency as a criterion for combination in image formation for two reasons. First, the mechanics of implementation of such an image are quite likely to be simplified under these conditions. The second reason is more complex; since each item of data is to appear but once in the file, the file can be organized on the basis of only one field, presumably that field on whose basis queries most frequently discriminate. With respect to the remaining data fields, therefore, the file is unorganized except insofar as these are correlated with the organizing field. Because of this, adjacency is as good a criterion as any other for combination in an image.

If n words are used to form a single image item, and if the main store contains M items, the image will consist of M/n items, each formed by combining logically n items in consecutive locations in the main store.

The simplest choice of function for use in the formation of such an image is that of logical "and" or "or." If we choose the latter function then we must search the items in the image for agreement with the "ones" of the query; unspecified fields of the query will be set to zero. Conversely, if we choose the former function, we must search the items in the image for agreement with the "zeros" of the query; unspecified fields of the query are then set to "one." Since these two are duals of one another, we will examine in detail only one of them--that employing "or." To repeat, the file will consist of two parts, a main store of M items and an image of the main store composed of M/n items, each formed by forming the bitwise logical "or" of n items in consecutive locations of the main store.

Examination of an entry in such an image cannot indicate the presence of an item having the desired characteristics--merely the absence of any item having these characteristics in the constituents of the image entry. With the above choice of function, the test performed on an image entry in the course of the search conducted on the image would be:

$$\text{(query)} \land \text{(image item)} = \text{(query)} \qquad \text{(bitwise)} \qquad (1)$$

Recall that unspecified data fields of the query are set to zero; this test therefore checks to see that the image item has ones in at least those positions where the query has ones. This is a necessary condition for one or more of the constituents of the image item to be desired items in the main store. Hopefully, the search conducted on the image will eliminate the necessity for searching several blocks of n adjacent words in the main file. The success or failure of the scheme will depend upon how many blocks are eliminated; thus it is necessary to attempt to make some estimate of this number.

To begin with, we note that the controlling factor for the number of successful responses to a search conducted on the basis of a given query will be the relative numbers of zeros and ones in the query, and in the file representation of the data.

We begin with some definitions and a model; let:

$q$ = total number of bits in an item of data as represented in the store

$m = \log_2$ (number of items of data)--i.e. $M = 2^m$

$r$ = number of ones in the specified fields of the query

$n$ = number of data words in the main store combined to form a single image word.

We must now postulate a distribution of the relative numbers of ones and zeros in the representations of the items of data. As a first hypothesis, we will assume that the $2^m$ representations of items of data are chosen at random from the $2^q$ possible q-tuples of ones and zeros. Further, we will assume initially that n = 2, and compute the probability that the bitwise logical "or" of two q-bit words so chosen will contain r ones:

total number of entries in "or" table for q-bit words $= 2^{2q}$

The logical "or" of two words will have a one in the $i^{th}$ position if either or both of its constituents have a one there. Therefore, any word with r ones can be formed as the logical "or" of two other words in exactly $3^r$ ways. Thus, the probability that the logical "or" of a randomly chosen pair of q-bit words will have ones in exactly r specified locations is $\dfrac{3^r}{4^q}$ .

A word will pass the test (1), however, if it has ones in <u>at least</u> the r locations specified by the query. The probability that the logical "or" of two randomly chosen words will have ones in at least r specified locations is

$$
\begin{aligned}
p_r &= \frac{3^r + \binom{q-r}{1} 3^{r+1} + \binom{q-r}{2} 3^{r+2} + \ldots + \binom{q-r}{q-r} 3^{r+q-r}}{4^q} \\
&= \frac{3^r \displaystyle\sum_{0}^{q-r} \binom{q-r}{i} 3^i}{4^q} \\
&= \left(\frac{3}{4}\right)^r
\end{aligned} \tag{2}
$$

Now we are in a position to estimate the number of comparisons we will have to make. If $p_r = \left(\frac{3}{4}\right)^r$ and if $t = (1 - p_r)$, then the probability of having k words pass test (1) in $2^m/n = 2^{m-1}$ trials is given by: $(p_r)_k = \binom{2^{m-1}}{k} (p_r)^k t^{(2^{m-1} - k)}$

Substituting for $p_r$ and t, we obtain:

$$(p_r)_k = \binom{2^{m-1}}{k} (\tfrac{3}{4})^{kr} \left[1 - (\tfrac{3}{4})^r\right]^{(2^{m-1}-k)} \tag{3}$$

The mean for this (binomial) distribution is:

$$\mu = 2^{m-1} (\tfrac{3}{4})^r \tag{4}$$

This means that if we search the file in two stages, first performing test (1) on each of the $2^{m-1}$ words in the image and then, when test (1) is successfully passed, examining each of the constituent words of the successful test word, then the expected number of successful test words will be $2^{m-1} (\tfrac{3}{4})^r$, where r is the number of ones in the query. Since each of these successful test words has two constituents, twice this many additional data words in the main store will have to be examined according to test (1) in order to complete the search. The total number of items which must be tested is therefore:

$$N_2 = 2^{m-1} + 2^m (\tfrac{3}{4})^r$$

$$= 2^m (\tfrac{1}{2} + (\tfrac{3}{4})^r)$$

The question now arises: What happens if we form the "or" of more than two adjacent words and let this serve as the image on which we search? By means of computations similar to those above it can be shown that, in general, the probability that the logical "or" of n randomly chosen q-bit words will have ones in at least r specified positions is $(\dfrac{2^n-1}{2^n})^r$. Under these conditions, it turns out that:

$$N_n = M (\tfrac{1}{n} + (1 - \tfrac{1}{2^n})^r) \tag{5}$$

Representing $N_n$ as follows:

$$N_n = (M) \times f(n, r)$$

we can plot the values of $f(n, r)$ for various values of n and r to determine, for a given r, what will be the optimal value of n.[2] A short program to do this was written for the IBM 7094, and values of $f(n, r)$ were computed for the range n = 2 to n = 15, r = 0 to r = 95. The results of this computation are depicted in Figure 1.

Before proceeding to examine these results in more detail, we pause to make the following observation: the greater r is, the smaller will be $f(n, r)$ and hence the more efficient the search procedure. Now if we have the capability of searching for zeros in the same way that we have just outlined that one would search for ones, then r need never be less than one-half the total length of all fields specified by the query, for if it is, we then conduct the dual of test (1) and represent the query by its zeros, setting the unspecified data fields to one. This requires, however, either that we have two logical elements--an "and" element and an "or" element-- attached to each bit position of each block of n words in the main store, or that we have a single element with two-function capability. Fortunately, a device with the latter property exists--the threshold device.[Ref. 2] Simply by altering the bias of the threshold device, one can change it from an "or" element to an "and" element and vice-versa; furthermore, the bias settings required for establishing these two modes of behavior are at opposite ends of the scale of possible bias settings and hence the discrimination or tolerance problems which have plagued threshold logic networks are as minimal as they can be. Such a memory system employing threshold elements for formation of the image should therefore have facility for program-controlled change of the threshold (or bias), and hence of the image-forming function, so that the latter may be chosen to match the characteristics of the current query.

Given, then, the fact that r need never be less than one-half the total length of specified fields of the query, we see that for this model, and for values of r in the range 13 to 38, the optimal value of n is 3, and that for the much larger range of 38 to 91, the optimal value of n is 4. We may also note that, over the latter

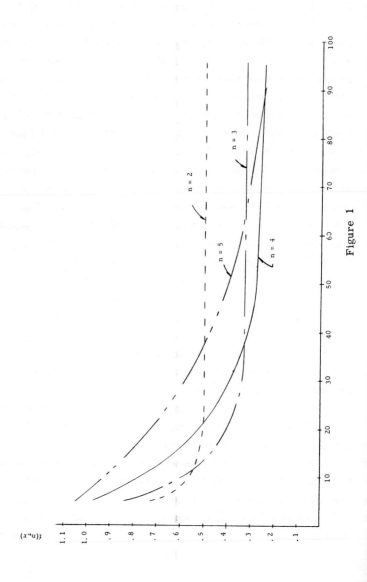

Figure 1

range, the value of $f(n, r)$ lies between 0.336 and 0.252, and for most of this range lies below 0.300.   This is a gratifying result, for it means a saving of a factor of between 3 and 4 in the number of items to be examined!  It is perhaps worthwhile at this point to note that this is the actual number of items examined: one need not scan the whole main store at any point.   One examines the items in the image, subjecting them to the test (1) (or its dual, depending on the nature of the query), and, when it is necessary, stops and immediately examines the constituent data words of an image word. (It is particularly easy to generate the addresses of constituent data words when n is a power of 2, as we shall see in a moment). Blocks of n words whose corresponding image word does not pass test (1) are never examined.

The fact that n = 4 is the optimal choice for a large range of reasonable values for data word size is particularly fortuitous because this means that the addresses of the initial words in successive image constituent blocks are congruent modulo $2^2$.   Therefore, if the initial word in the image is numbered 0 and has as constituents data words in the main store in locations 0 to 3, the addresses of the constituents of the image word numbered Y (in binary) will be Y00, Y01, Y10, Y11.

III.    Summary of System Characteristics

A brief summary of the proposed system is, then, the following:

1)   There is a large main store where the actual data is stored; for purposes of the kind of search proposed, only serial access will be required, although random access may be desired for other reasons.   Similarly, nothing need be specified about read-write capability for our purposes.

2)   There is an image store, one-quarter the size of the main store.   Each word of the image is formed as a bitwise logical function of 4 words in the main store.   The logical devices used to implement these functions are threshold devices, and the bias of these devices may be altered under program control, changing them

from "and" to "or" elements.

3)  Searches are conducted in such a device in the following way:  One begins searching on the image store; when one encounters an image item which passes the test, and hence may number among its constituents one or more data items which are sought, one then proceeds to generate the addresses of these constituents and to examine them individually.

4)  With such a device and procedure, and under the very crucial assumption that the items of data are chosen at random from among the set of all possible data words, one can expect, on the average, to have to examine a total number of items which is one-quarter to one-third the number of items in the main store and thereby retrieve all relevant data items.

In this range, then, an increase in equipment of about 25 per cent seems to yield a gain in efficiency of a factor of between 3 and 4.

IV.  False Drops and Correct Responses

It is reasonable to inquire at this point what the chances are, with such a scheme, of generating "false drops." That is, if an image item passes test (1) or its dual successfully, what is the probability that at least one of its constituents is a desired data item?

We begin by finding the probability, $h'_r$, that the image word constructed as the bitwise logical "or" of n randomly chosen words will have ones in exactly r specified locations and that at least one of its constituent words will be a desired word (have ones in the r specified locations).

Suppose k words have ones in all r positions; then (n-k) words have ones in fewer than r positions.  We may now ask, in how many ways can this happen?  Each of the (n-k) words can be in any of $(2^r-1)$ possible states; therefore, there are $(2^r-1)^{n-k}$ ways of assigning ones to the remaining (n-k) words in such a way that none has r ones.  There are thus $\binom{n}{k} (2^r-1)^{n-k}$ ways of assigning ones to n words in such a way that k words have ones in all r specified positions and the remaining (n-k) words do not. Thus, the probability, $h_r'$, that the image word constructed as the "or" of n randomly chosen words will have ones in exactly r specified locations and that at least one of its constituents will be a desired word is:

$$h_r' = \frac{\sum_1^n \binom{n}{k} (2^r-1)^{n-k}}{2^{nq}} = \frac{(2^r-1)^n \sum_1^n \binom{n}{k} (2^r-1)^{-k}}{2^{nq}}$$

$$= \frac{(2^r-1)^n \left[ (1 + (2^r-1)^{-1})^n - 1 \right]}{2^{nq}}$$

$$h_r' = \frac{2^{rn} - (2^r-1)^n}{2^{nq}}$$

From this, as before, we can compute the probability that the image word constructed as the logical "or" of n randomly chosen words will have ones in at least r specified locations and that at least one of its constituent words will be a desired word. As before, for each of the acceptable (in the above sense) configurations in the r specified positions, there are $2^{n(q-r)}$ acceptable configurations in the (q-r) unspecified positions. Therefore, the probability, $h_r$, that the image word constructed as the logical "or" of n randomly chosen words will have ones in at least r specified positions and that at least one of its constituent words will be a desired word is:

$$h_r = \frac{(2^{rn} - (2^r-1)^n) \, 2^{n(q-r)}}{2^{nq}} = \frac{2^{rn} - (2^r-1)^n}{2^{rn}}$$

$$h_r = 1 - \left( \frac{2^r-1}{2^r} \right)^n \tag{6}$$

Therefore, $d_r$, the probability of a false drop is:

$$d_r = \left( \frac{2^r-1}{2^r} \right)^n \tag{7}$$

Unfortunately, but not surprisingly, this last probability has

a tendency to increase with r.    Thus, the greater the number of
ones in the query, the more likely it is that the constituents of an
image word which passes test (1) will not be desired words.    There
is a mitigating factor, however, in that the absolute number of drops
is itself small; therefore, even if a large percentage of these are
false drops, the number of such remains small.

The probability can be used to find the expected number of
correct responses to a query of r ones.    Since the search is (ultimately) a matter of a choice between two alternatives, the distribution is binomial and has an expected value of

$$\frac{M}{n} \left( 1 - \left( \frac{2^r - 1}{2^r} \right)^n \right)$$

This figure can be used to help evaluate the assumption of randomness made here with respect to any actual proposed data base.

## V.    A Modified Search Strategy

Thus far, computations have been carried out in terms of the
number of ones in specified fields of the query, or of the number of
zeros in specified fields of the query.    In terms of these parameters, expressions were derived for the probability that an image
item would pass text (1), for the expected number of items to be
examined in a search conducted on the basis of a query word, for
the probability of a false drop, and for the expected number of correct responses.    In the course of this development, we observed
that r, the number of ones (or zeros) in the specified fields of the
query need never be less than one-half the total length of these
fields.    We now observe, however, that since we are already advo-
cating a two-function capability, we can arrange to conduct both test
(1) and its dual on image items which pass test (1), and thus further
reduce the expected number of successful (but undesired) responses.
In particular, such a scheme has the effect of increasing r to equal
the total length of specified fields of the query, at the cost of examining some items in the image twice.    We will compute the expected
number of items to be examined under these conditions, but first it

should be observed that such a scheme has practical limitations. Although such a change in image function requires merely a change in bias, the large number of logical elements connected to the bias constitutes a very large capacitance and hence ultimately restricts the speed of voltage changes.   Therefore, program control of such changes is likely to slow down the search if exercised too frequently.

Let us now review this proposed technique:  We must have both the "standard" and "dual" forms of the query available.   We proceed as before through the image, subjecting each image item to test (1).   When test (1) is passed by an image item, we change the image function to its dual, and test the dual image item against the dual query.   We then change the image function back to its more efficient form and, depending on the outcome of the second test, either go on to the next image item or examine the constituents of the present item for equivalence with the specified fields of the query and then go on to the next image item.

Since the test (1) and its dual are performed independently, one can multiply their individual probabilities to find the joint probability.   Thus, the probability that an image item whose constituents are randomly chosen q-tuples will pass both test (1) and its dual for a query with r ones out of a total specified field length of    is

$$\left( \frac{2^n - 1}{2^n} \right)^{\ell}$$

.   This enables us to compute $N_n^*$, the expected number of items to be examined in any search,  counting dual tests of the same item as two examinations.   The expected number of items which will pass test (1) in $\frac{M}{n}$ trials is, as before

$$\left( \frac{M}{n} \right) \left( \frac{2^n - 1}{2^n} \right)^{r}$$

.   This many items will require dual testing.

The expected number of items which will pass both test (1) and its dual is

$$\left( \frac{M}{n} \right) \left( \frac{2^n - 1}{2^n} \right)$$

.   n times this many constituent items in

the main store will have to be examined.   Thus

$$N_n^* = M \left( \frac{1}{n} + \left( \frac{2^n-1}{2^n} \right)^\ell + \frac{1}{n} \left( \frac{2^n-1}{2^n} \right)^r \right) \qquad (8)$$

We now inquire under what conditions the second scheme for searching will be more efficient than the first. We require:

$$\overset{N_n}{M \left[ \frac{1}{n} + \left( \frac{2^n-1}{2^n} \right)^r \right]} > \overset{N_n^*}{M \left[ \frac{1}{n} + \left( \frac{2^n-1}{2^n} \right)^\ell + \frac{1}{n} \left( \frac{2^n-1}{2^n} \right)^r \right]}$$

$$\left( \frac{2^n-1}{2^n} \right)^r > \left( \frac{2^n-1}{2^n} \right)^\ell + \frac{1}{n} \left( \frac{2^n-1}{2^n} \right)^r$$

$$\left( 1 - \frac{1}{n} \right) > \left( \frac{2^n-1}{2^n} \right)^{\ell-r} \qquad (9)$$

When n = 4, condition (9) becomes:

$$\left( \frac{3}{4} \right) > \left( \frac{15}{16} \right)^{\ell-r}$$

$$\ln \left( \frac{3}{4} \right) > (\ell - r) \ln \left( \frac{15}{16} \right)$$

$$(-0.287) > (\ell - r)(-0.06454)$$
$$(\ell - r) > 4.4 \qquad (10)$$

Therefore, the latter system is more efficient than the former when n = 4 for queries in which the difference between the total number of ones in the specified fields of the query and the total number of zeros in the specified fields of the query is greater than 4.

## VI.  A Second Summary

We may summarize the salient characteristics of the proposed system in the following way:

1)  An increase by a fraction 1/n in equipment yields an increase of slightly less than a factor of n in search efficiency.

2)  The probability of false drops is high, but the probability

of any drop is low, and so the number of false drops is low.

3)   An application of threshold logic is suggested which is singular in two respects:

a)   the problems of discrimination in level which plague such devices are minimized

b)   the multi-function capability of threshold devices is utilized.

4)   Although n = 4 seems to be a reasonable value for most applications, longer data words would enable one to use larger n and thus construct an even more efficient system.

VII.   Results of Simulation

In an effort to evaluate the validity of the assumptions made here, and hence of the conclusions drawn, the system proposed here was simulated on the IBM 7094, using selected portions of an actual personnel file of 2156 items.[3] Each item in this experimental file was 144 bits in length and consisted of the following kinds of information: 3 dates, a department number, an occupation code, an education code   together with a major subject code, an initial, a project number, and a two character "supplemental code."

The items in the main file from which the experimental file was drawn were ordered on the basis of man number, although the man numbers were not present in the experimental file.   Thus, the organization of the file conformed quite precisely to the hypothesis of small correlation between adjacent file entries.   There was, however, extensive correlation between fields within a single entry; persons of given educational level tended to have the same occupations, for example, and persons born after a certain date had ended their education at the high-school level.   Also, certain projects and departments were staffed mostly by physicists, others by mathematicians, etc.   In view of these rather extensive departures from the hypothesis of no correlation inherent in the assumption of random choice of n-tuples, it is not surprising that none of the experimental sample points lie within 2 standard deviations of the expected values predicted by the simple model.   What is reassuring, however,

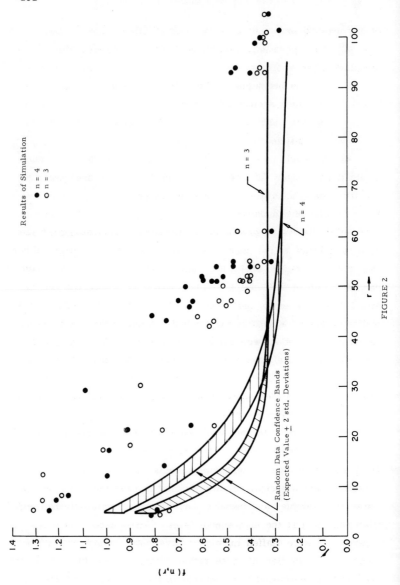

FIGURE 2

is that the qualitative behavior predicted by the model is clearly reflected in the data.    (The asymptotic limits are, of course, necessarily the same.)   Furthermore, in all but three cases of values of
r of 15 or more, the number of searches necessary in the experimental file with the proposed organization was smaller than would
have been required without it.    The results of the simulation are
shown in Figure 2. [4]

VIII.   Conclusions

The question of validity of the assumption of randomness of
the data base is a central one.   It is easy to conceive of situations
in which the assumption is justified.   For example, in a motor vehicle license file arranged either by license number or alphabetical-
ly, it is unlikely that much correlation exists between adjacent entries with respect to year, make, color, model, etc.   As we have
seen, however, even in a personnel file where the correlation between certain fields (such as name, birth date, etc.) is likely to be
quite small, the correlation between other fields can be quite large.

It does not seem reasonable to attempt to formulate a model
for a data base with cross-correlation between fields.   The expressions for expected numbers of searches in such a model would very
likely be so highly parameterized and so complex as to preclude
useful insight.   It does seem probable that the efficiency of the system proposed here could be either enhanced or impaired, depending
upon the basic nature of the data, the encoding representation chosen, and even on the query made.

Two further remarks are in order with regard to implementation of such a system.   It has been assumed throughout this inquiry
that the hardware for forming the image is duplicated for each block
of n items in the main store; if a serial access device were used
as the main store, however, it is entirely possible that one might
form the image items one-by-one as successive blocks of n data
words came past a certain "image station."   The image-forming
hardware would then exist but once, and the image itself would ex-

ist only one item at a time.  A successful response to the test at
the "image station" might then trigger a detailed search of the constituents of the appropriate item at another reading station further
on in order to prevent one's having to wait for the constituents to
"come around again. "

A third approach would be to store the image in a smaller,
fast-access memory, while maintaining the main file in a larger,
cheaper, slower-access store.  In such an implementation, the net
gain in speed would be even greater.  It would be quite easy (as
was done in the simulation discussed previously) to form the image
by means of a subroutine at storage time.

The second remark which should be made is that although the
amount of hardware to be associated with the store is much greater
than the amount to which one has become accustomed, the purposes
of this system are somewhat different from those of a general-purpose machine.  Furthermore, by distributing the hardware over the
memory one can arrange for a simpler central processor, and thus
the net cost may not be increased appreciably.  It is this writer's
feeling that time--and maybe even theory--will prove that a more
"diffuse" organization of processors is desirable in information retrieval problems.

As to the future, further investigation into the effect sof data
correlation is planned, as mentioned.  It would also be desirable to
experiment with other functional images of the data.  The present
image-function (pair) seems reasonable from many standpoints, but
it is by no means clear that others may not be better.  Further investigation is planned in this area also.  Finally, some additional
simulation might be undertaken to test further the validity of both
system and assumptions.

IX.  Acknowledgement

Thanks are due D. L. Reich for assistance in programming
and a critical reading of the derivations.

## Notes

(1)   The work reported here was supported under Contract AF-19(628)-2752.

(2)   Attempting to find the optimal values by looking for zeros of  $\frac{\partial}{\partial n}$  f(n, r) is most unrewarding, for:

$$\frac{\partial}{\partial n} f(n, r) = - \frac{1}{n^2} + r \left(1 - \frac{1}{2^n}\right)^{r-1} \left(\frac{\ln 2}{2^n}\right)$$

(3)   2154 items for n = 3.

(4)   Note that the expected number of searches and the standard deviation cannot simultaneously be normalized to be independent of M.   The confidence bands plotted are, therefore, valid only for a file of the same size as the experimental file.

## References

1.     Becker, J. and Hayes, R. M.; Information Retrieval: Tools, Elements, Theories; New York, John Wiley; 1963.

2.     Muroga, S.; "Logic Elements on Majority Decision Principles and Complexity of Their Circuits"; ICIP, UNESCO; June, 1959.

3.     Reich, D. L.; "Associative Memories and Information Retrieval." This volume, III F.

# III E.  File Memory Addressing[1]

### J.  Beatty[2]
### and
### S.  Muroga[3]

## 1.  Introduction

In this report we discuss how to address a file memory and how to construct it such that addressing is economical and fast.  In other words, a large number of pieces of information (e. g. personnel records) are stored in a memory and when some digits of a piece of information (e. g. the name of an employee) are given as an identifier (or a key), we want to locate information associated with the given identifier in memory.  Access to the information should be very fast.  Easily conceivable approaches like table-look-up or binary search are too slow.

A question arises whether we can set up some transformation between a key and an address and to store all pieces of information in a memory such that this transformation will work.  Buchholz surveyed [(1)] several proposed methods along this line, with different types of transformation.  However, there are two different motivations behind them.  One is that we want a one-to-one correspondence between keys and addresses, as Buchholz recommends.  The other is that we want to convert keys into addresses which are as random as possible, since a one-to-one correspondence is hardly to be expected in most cases.  The primary objective of randomization is as follows: some keys may be transformed into the same address, so we must search for the right key in a group of keys with this address.  By randomization, we want to make the number of such keys

206

for each address as uniform as possible, hoping not to make search time in a certain group of keys too long.

## 2.  Our Approach

One-to-one correspondence is certainly advantageous (because we can reach the right key immediately), but unless we look at a given set of keys in advance and set up a transformation suitable for it, we are not guaranteed one-to-one correspondence.  Our approach is to look at a given set of M keys in advance and to find N Boolean functions of the digits of a key (regarded as a binary number) such that the sequence of N binary digits (this sequence will be used as the address to the memory), each of which is a Boolean function of the digits of a key, is different for each key in the given set. (This gives a one-to-one correspondence between keys and addresses.) Here N is the least integer not smaller than $\log_2 M$.

When N such Boolean functions are found, we implement these functions.  (If a fast access memory is used as a file memory, a circuit with hardware which performs these functions should be implemented.  If a slow access memory is used, expressions for these functions may be stored in  memory and functions may be calculated with software.)  When we want to use this file memory for a key given from outside, an address will be computed as functions of digits of the key and information associated with the key is reached in the file memory.

Discovery of a good set of functions by examining a given set of keys may be time-consuming but once it is discovered, addressing to a file memory is done very quickly and with a minimum number of unused memory locations.

When a set of given keys is fixed, a once obtained transformation (i. e. functions) can be used until we make a change in the set of keys. (Note that changes in information but not in keys require no change of transformation.) However, when keys are changed often, finding a new transformation each time is too cumbersome. So a multi-occupancy file memory would be more desirable and a transformation for this organization is to be discovered. In other words, the memory is divided into groups of memory locations. Each group is called a bucket and each bucket has only one address. Each bucket includes pieces of information for more than one key and possibly some empty memory locations. Keys for information in the same bucket are transformed into the address of the bucket. If a new key is added, information for it is stored in an empty memory location in a bucket whose address is one transformed from the key, or in the first empty memory location in subsequent buckets. So when a key is given from outside for inquiry, we must search through the subsequent buckets starting with the bucket addressed, until the right information is reached.

A transformation which makes the number of keys in each bucket uniform is desirable in order to minimize the search time. Even though there is nonuniformity in bucket size to some extent, the above multi-occupancy memory works fairly well. So if we cannot discover a transformation giving one-to-one correspondence in a certain computation time, we can use the multi-occupancy memory, based on Boolean functions. If some Boolean functions are dropped from N functions which are a best set obtained in a specified computation time, these functions will be a good transformation which will give a reasonably uniform bucket size.

## 3.   Problems in Our Approach

The following problems are of vital importance to determine if our approach is practically useful. The first problem is to estimate computation time to find a transformation, by examining a given set of keys. If computation time is too great, our approach would be useless. The second problem is complexity of the transformation. If it is too complex, implementation of a circuit or storing functional forms would be difficult and when the memory with the obtained transformation is in operation, computation of an address from a given key may take too much time. Thus, our approach would not be good when the number of keys is very large.

We encounter the addressing problem not only in the RAMAC disc memory but also in various environments under different motivations and restrictions.

Our programming experiments made so far may not be sufficient to get definite conclusions but seem to imply usefulness of our
approach, at least for a small number of keys.

### 4.  Formulation and Outlines of Algorithms

Our approach may be formulated as a synthesis problem of a
multi-output combinational network of a certain type, if we want one-to-
one correspondence between keys and addresses.    Arrange vertical-
ly M keys consisting of binary numbers and regard this matrix as
a truth table where columns represent input variables.    Let us define $\mu(x) = [\log_2(x)]$ = (the smallest integer j such that $x \leq 2^j$).
Then we want to associate a different binary number less than
$2^{\mu(M)}$ to each row of the table.    This synthesis problem has the
following two properties:    (1)  There is a large number of don't
care conditions (because the length of a key is about 35 to 70 bits
and the number of keys is much smaller than $2^{35}$).    (2)  The binary
numbers obtained as outputs are not required to have any particular
order but only to be distinct.    Even if this second characteristic is
dropped, we can always synthesize the network since it is well
known that if there is no duplication in the M rows of the truth table, it is always possible to synthesize outputs such that different
binary numbers less than $\mu(M)$ are associated with rows in a
specified way.    However, the standard procedures for doing this do
not make use of property 2.    It is hoped that this property can be
taken advantage of.

Any Boolean function $f(x_1, x_2, \ldots x_n)$ can be expanded into
polynomial form using modulo two sum + ;

$$f(x_1, x_2, \ldots x_n) = g_0 \oplus g_1 x_1 \oplus g_2 x_2 \oplus \cdots \oplus g_n x_n$$
$$\oplus g_{12} x_1 x_2 \oplus g_{13} x_1 x_3 \oplus \cdots \oplus$$
$$\oplus g_{n-1, n} x_{n-1} x_n$$
$$\vdots$$
$$\oplus g_{12 \ldots n} x_1 x_2 \cdots x_n \, ,$$

where $x_1, x_2, \ldots x_n$ are columns of the truth table, i.e. the digit positions of a key and where $g_i$ are coefficients which are either 1 or 0. Let us construct the outputs, using the above polynomial expansion.

The approach we have tentatively adopted is to start by selecting Boolean functions $f_1, \ldots f_{\mu(M)}$ (of n variables each) from a certain class of Boolean functions C (e.g. the class of linear functions in the sense of modulo 2 addition) which give us as uniform a partition of the keys as possible and then to modify the $f_j$ by addition of correcting terms until all keys are transformed to distinct addresses. A more precise statement is as follows:

If X is an n-bit vector whose coordinates are 1 or 0, we write $f(X) = (f_1(X), \ldots, f_{\mu(M)}(X))$. Then f transforms keys X into addresses $f(X)$. We denote by $\pi_f$ the partition of the set of keys defined as follows: keys X and Y are in the same block of $\pi_f$ if and only if $f(X) = f(Y)$. Our ultimate aim is to choose f so that $\pi_f$ is the 0 partition, i.e. all blocks of $\pi_f$ consist of single keys. An intermediate aim is to find an f so that the sizes of blocks of $\pi_f$ for a certain specified number of functions, $v < \mu(M)$, are as uniform as possible.

Assuming that we have a function f, the following algorithm shows how one can refine f to obtain the desired separation of keys:

Algorithm 1. Let $w(X)$ be the number of ones in the vector X (i.e. a key). Order the keys and number them as $X^i$ with an index i, so that $w(X^i) \leq w(X^j)$ if $i < j$. Let g be the smallest number such that $f(X^g) = f(X^h)$ for some $i < g$, i.e. for $h < g$. (If there is no such number, then f already separates the keys). f yields an identical address $f(X^g)$ for keys $X^h$ and $X^g$. Let $x_{g_1}, \ldots, x_{g_k}$ be the positions of $X^g$ having ones among $x_1, \ldots, x_n$ and let $A = (a_1, \ldots, a_{\mu(M)})$ be any binary number not appearing in the keys $f(X^i)$ for all $i < g$. Form the vector $B = A \oplus f(X^g)$, where modulo two addition $\oplus$ is done coordinate by coordinate. Let q be the conjunction of $x_{g_1}, \ldots, x_{g_k}$. Then modify f by adding modulo two the vector qB to all the keys $f(X^i)$, i.e. by adding modulo two $qb_m$ to $f_m$ for every m such that $1 \leq m \leq \mu(M)$. Clearly this insures that $f(X^g) \neq f(X^i)$ for all $i < g$.

Next find the next smallest number g! such that $f(X^{g!}) \neq f(X^{h'})$ for $h' < g'$. Repeat the above process.

The process is repeated until the keys are separated.

(The case of forming addresses with only the first degree terms in the abovementioned polynomial expansion has an interesting relation with a coset expansion of an Abelian group.[2] Some work[8] published in the past proposed use of the first terms only but fixed forms in adding the first degree terms were used for any set of keys. Note the difference of this from our approach in which functional form is to be searched for a given set of keys. Application of error-correcting codes[4] [5] [7] also used the fixed functional form of the first degree terms).

Since the second or higher degree terms give finer adjustment than the first degree terms would do, the first degree terms are supposed to give the first-order approximation in separating the keys by distinct addresses. So the following also may be reasonable.

Algorithm 2. After obtaining a linear function f, i.e. using only the first degree terms, modify f as follows to obtain a complete separation. Suppose $f(X^i) = f(X^j)$ for a certain i and suppose $A = (a_1, \ldots a_{\mu(M)}) \neq f(X^p)$ for every p such that $1 \leq p \leq M$. Let $B = A \oplus f(X^i)$ and let $y_k = x_k$ or $\bar{x}_k$ according as $x_k{}^i$ (the kth coordinate of $X^i$) = 1 or 0. Now modify f by adding (mod 2) the vector $y_1 \ldots y_n B$ to every $f(X^p)$. This clearly leaves $f(X^p)$ unchanged for $p \neq i$ and converts $f(X^i)$ into A.

Given a class C of candidate Boolean functions, the problem of finding a vector function $f = (f_1, \ldots f_m)$ which separates the keys with minimum m and $f_j \in C$ for $1 \leq j \leq m$ can be formulated in integer linear programming as follows:

Let $b_{ij} = f_j(X^i)$, $f_j \in C$ and $1 \leq i \leq M$. Let $a_{ij} = b_{ij} \oplus b_{i'j}$ where (i, i') ranges over all ordered pairs of integers $1 \leq i < i' \leq M$. Now we want to minimize $\sum_{i=1}^{t} x_i$

subject to the constraints that $x_j = 0$ or 1 and

$\sum_{j=1}^{t} a_{ii'j} \cdot x_j \geq 1$, for $1 \leq i < i' \leq M$, where t is the num-

ber of candidate functions in C. The number of inequalities is at most $M^2/2$, so this approach may be useful only for a small number of keys. If we can know an approximate solution for a large number of keys, this approach may be used as an auxiliary means to complete the solution. This type of linear programming problem is a special case of the minimum cover problem of graph theory. [4]

In order to decrease the number of variables necessary in the previous algorithms one could find a function $f = (f_1, \ldots, f\alpha)$, $f\alpha \in C$ such that f separates the keys with $\alpha$ greater than $\mu(M)$. Then if $\alpha > \mu(M)$, algorithms (1) or (2) could be applied to the transformed keys $f(X^i)$. (not only to the original keys, $x^i$). An algorithm for this has been programmed and is described in Section 5.

Whatever algorithm we follow, it is impossible to exhaust all possible Boolean functions and choose appropriate ones which completely separate keys. K. Appel of the University of Illinois suggested the use of functions which have equal or almost equal number of ones and zeros, as candidate functions. This is because if M is equal to some power of 2, the number of ones and zeros in each of $\mu(M)$ columns separating keys completely is equal.

5.   Description of the Program

Let us describe what algorithm we actually programmed and how it is programmed in order to keep computation time as small as possible. Let us designate the M x n matrix of keys by K, whose ith row is the key $k_i$ and jth column is $x_j$. Let S(K) be the set of all linear functions; $g_1x_1 \oplus \cdots \oplus g_nx_n$.

As the first step, instead of trying to get $\mu(M)$ columns of complete separation, we tried to find a minimal subset $\{\sigma_1, \ldots, \sigma_a\}$ of S(K) such that no two rows are equal in the M x a matrix A whose ith column is $\sigma_i$. (Note that if the ith row of A is taken as the address corresponding to the key $k_i$, this gives the solution of parity-check addressing. [2])

Where $\sigma_j \in S(K)$ for $1 \leq j \leq m$, define $D(\sigma_1, \ldots, \sigma_m)$ as the M x m matrix whose ith column is $\sigma_i$ and $\pi(\sigma_1, \ldots, \sigma_m)$ as the partition of the rows $\rho_i$ of $D(\sigma_1, \ldots, \sigma_m)$ determined by the equivalence relation $\rho_i = \rho_j$. Thus the blocks of $\pi(\sigma_1, \ldots, \sigma_m)$ are the sets of equal rows of $D(\sigma_1, \ldots, \sigma_m)$. $\{\sigma_1, \ldots, \sigma_m\}$ will be called a splitting set if and only if no two rows of $D(\sigma_1, \ldots, \sigma_m)$ are equal. Let $\phi(\sigma_1, \ldots, \sigma_m)$ be the sum of the squares of the sizes (i.e. numbers of rows) of the blocks of $\pi(\sigma_1, \ldots, \sigma_m)$. Now Appel's suggestion is to choose from among a random selection of elements of $S(K)$ an element $\sigma_1$ such that $\phi(\sigma_1)$ is minimal. Now suppose $\sigma_1, \ldots, \sigma_p$ has been chosen. Then one selects $\sigma_{p+1}$ from among a random selection of elements of $S(K)$ such that $\phi(\sigma_1, \ldots, \sigma_{p+1})$ is minimal. This process is continued until one has $\phi(\sigma_1, \ldots, \sigma_p) = M$ (which is a necessary and sufficient condition that $\{\sigma_1, \ldots, \sigma_p\}$ is a splitting set). Let us call a splitting set $\{\sigma_1, \ldots, \sigma_p\}$ optimal if and only if $2^{p-1} < M \leq 2^p$. The purpose of the criterion function here is to guarantee that the successive partitions obtained will have uniform block sizes. Clearly, if it is possible to obtain the absolute minimum for $\phi$ at each step (i.e. to choose $\sigma_1, \ldots, \sigma_p$ such that the blocks of $\pi(\sigma_1, \ldots, \sigma_p)$ differ in size by at most one), then the procedure will yield an optimal set. In programming Appel's procedure for the IBM 7094, one runs into the difficulty that the generation of elements of $S(K)$ and the computation of $\pi(\sigma_1, \ldots, \sigma_p)$, in order for both to be efficient, requires the data to be stored in essentially different formats. This was circumvented by initially generating a number of elements of $S(K)$ more or less at random and then, after a conversion to a different format, selecting $\sigma_j$ from these elements by the method described above.

Phase 1: Generation of Elements of $S(K)$

Let us define $\psi(\sigma_i)$ as the absolute difference between the number of ones and the number of zeros in $\sigma_i$. In case $M=2^\alpha$ with some integer $\alpha$, a set $\{\sigma_1, \ldots, \sigma_p\}$ will be optimal only if $m=\alpha$ and $\psi(\sigma_i)=0$ for $1 \leq i \leq p$. If M is not a power of 2, then it is no longer necessary that $\psi(\sigma_i) = 0$ but it appears likely that the numbers $\psi(\sigma_i)$ would have to be small.  (See the results in the appendix.)  On the basis of this, an element $\sigma_i$ of S(K), after being randomly generated, is accepted or rejected according as $\psi(\sigma_i) \leq \varepsilon$ or not, where $\varepsilon$ is a preassigned number.  Let the set of these accepted $\sigma_i$ be T.

Phase 2:   Search for a Solution

Appel's procedure is now applied to the set T of acceptable elements of S(K) provided by Phase 1.

In place of the criterion function $\phi$, phase 2 was also tried with a function $\phi'$ defined as follows:   $\phi'(\sigma_1, \ldots \sigma_m) = 2^{18} \cdot x + y$ where x is the size of a largest block of the partition $\pi(\sigma_1, \ldots \sigma_m)$ and y is the number of such largest blocks.

6.   Experimental Results

As yet the program has been tested on randomly generated data with the IBM 7094.   The preliminary conclusions that may be drawn are as follows:

1.   $\phi$ seems to be slightly superior to $\phi'$ as a criterion function on the average.

2.   The program handles up to 8000 keys of 36 bits each, and can be modified to accommodate this many 108-bit keys.

3.   Increasing the size of the set T generated by phase 1 seems to improve the performance of phase 2 slightly, though it gives rise to a roughly linear increase in computation time.

4.   Using values of $\varepsilon$ in phase 1 that are small relative to the number of keys seems to improve the performance of phase 2 slightly.

5.   For sets of 100 keys, minimum splitting sets of 8-10 columns were obtained.   For 500 keys, 14-15 columns were required. As we have more columns, the size of the blocks gets smaller and we completely separate keys at the tenth column.

6.   Computation times are less than a minute for up to 5000 keys when T has 36 elements.

Appendix:   Miscellaneous Theoretical Results

In an attempt to characterize splitting sets, the following was observed:

Theorem 1:   Let $M = 2^{\alpha}$ and let $\sigma_j$ be an M-dimensional Boolean (column) vector for $1 \leq j \leq \alpha$.  Then $\{\sigma_1, \sigma_2, \ldots, \sigma_\alpha\}$ is a splitting set, if and only if $\psi(\sigma) = 0$ for each $\sigma = a_1 \sigma_1 \oplus \ldots \oplus a_\alpha \sigma_\alpha (a_j = 0$ or $1)$ with not all $a_j = 0$.

This theorem shows the converse of a special case of D. E. Muller's famous theorem[Note 5] on coding.

Corollary 2:   Let $\sigma_1, \ldots, \sigma_\alpha$ be M-dimensional Boolean vectors such that $2^{\alpha-1} < M \leq 2^\alpha$ and $\{\sigma_1, \ldots, \sigma_\alpha\}$ is a splitting set. Then if $\sigma = a_1 \sigma_1 \oplus \ldots \oplus a_\alpha \sigma_\alpha$ and not every $a_j = 0$, then $\psi(\sigma) \leq 2^\alpha - M$.

Corollary 3:   Let K be an M x $\alpha$ Boolean matrix where $M = q \cdot 2^\alpha$ with columns $\sigma_j$.  Then the blocks of $\pi(K)$ each have size q (i.e. the number of rows in each block) if and only if for any $\sigma = a_1 \sigma_1 \oplus \ldots \oplus a_\alpha \sigma_\alpha$ with not every $a_j = 0$ we have $\psi(\sigma) = 0$.

Theorem 4:   Let K be an M x $\alpha$ Boolean matrix with distinct rows. Let K' consist of any m of the columns of K.   Then the size of every block of $\pi(K')$ is $\leq 2^{\alpha-m}$.

## Notes

1.   Status Report for Air Force Cambridge Contract AF19-(628)-2752.

2.   IBM Research Division, Yorktown Heights, N. Y.

3.   Formerly, with IBM Research Division, Yorktown Heights, N. Y.   Currently, with the Department of Computer Science, University of Illinois, Urbana, Ill.

4.   The problem of minimum prime-implicants expression studied by Cobham, McCluskey, Petrick and others is somewhat similar to our problem.   Ray-Chaudhuri has proposed an approach (9) to the minimum cover problem which is not based on linear programming.

5.   D. E. Muller:   "Application of Boolean Algebra to Switching Circuit Design and Error Detection. "   IRE Trans., EC-3 (1954,

pp. 6-12.

## References

(1)  W. Buchholz:  File Organization and Addressing, IBM System Journal, June 1963, pp. 86-111.

(2)  S. Muroga:  Application of Group Theory and Coding Theory to File Memory Addressing, IBM Research Report RC 1025, Sept. 1963.  Abstract was presented at International Conference on Microwaves, Circuit Theory and Information Theory, 1964.

(3)  A. D. Lin:  Key Addressing of Random Access Memories by Radix Transformation, 1963 Spring Joint Comp. Conf., pp. 355-366.

(4)  A. Schay and N. Raver:  A Method for Key-to-Address Transformation, IBM Journal, Apr. 1963.

(5) M. Hanan and F. P. Palermo:  An Application of Coding Theory to a File-Addressing Problem, IBM Journal, Apr. 1963.

(6)  W. W. Peterson:  Addressing for Random-Access Storage, IBM Journal, Apr. 1957.

(7)  C. V. Freiman and R. T. Chien:  Further Results in Polynomial Addressing, IBM Journal, Oct. 1963.

(8)  A. C. Reynolds, transl.:  Patent No. 871,256 in England, 1961.

(9)  D. K. Ray-Chaudhuri, "An Algorithm for a Minimum Cover of an Abstract Complex," Canadian Journal of Math., Vol. 15, pp. 11-24, 1963.

III F.    Associative Memories and Information Retrieval[1]

D.  L.  Reich

The study and practice of information retrieval is usually concerned with methods for extracting specified parts from large collections of stored data.    When these methods involve computers,
special attention must be paid to the design and use of the storage
media of the machine.    The hardware and algorithms are generally
expected to be able to provide fast, efficient searches of large
blocks of stored data or, alternatively, to do rapid path-tracing
through stored relationships until the desired item is found.    Sug-
gestions for solutions to these problems have appeared in the technical literature under the heading "associative memories."
　　Three basic properties ascribed to proposals with this name
are:

A.    Linkage of stored items on the basis of content.
B.    Parallel search capability over all or part of the memory
        at a given time.
C.    Distributed rather than centralized control for parallel
        processing.

These three properties are independent: none requires another for
its realization.

A.    Linkage of Stored Items by Programming

　　The first property is usually described in software terms such
as lists, list structures, threaded lists, end-linked lists, doubly
linked lists or multi-lists.    (The IPL languages form the basis of
the work on lists and list structures.)    Data stored on an IPL-type
list have an intrinsic partial ordering placed on them.    The main

217

feature of this ordering is the following. It is simple to get from any item to any item below it on the list structure, and it is difficult to get to any item above it or not related to it on the structure. Such a structure may be used for storing an inventory list by system, system component, component part and so on down. Information would then be retrieved by a search down the proper branches of the structure.

Perlis' "threaded" list concept changes the intrinsic partial ordering to a total ordering over the entire list structure. Each element in the structure contains some status bits in addition to the normal contents and link bits. This extra storage space is used to indicate whether an element is the head of a sublist, end of a sublist or normal element of a sublist. The link part of the last element in a sublist points to the place in the main list at which the sublist branched. The result of this technique is elimination of much of the "housekeeping" associated with pushdown lists: the entry point for a sublist is permanently stored in its tail and need not be recorded each time the sublist is entered.

Many information retrieval problems do not involve data sets with order relations on them. End-linked lists provide the capability for easily accessing any item in a data set from any other in the same set. The last item in the list is linked to the first item. Iteration over the whole set is then quite simple regardless of starting point. In distinction to threaded lists, no status bits are required but the normal pushdown techniques must be followed. Doubly-linked lists are a refinement of end-linked lists providing a link up as well as down the list for every element. This convenience is paid for through the use of more storage space.

Multi-lists involve the use of descriptors, coordinate indexing and, as the name implies, many inter-connected lists to provide the usual storage and retrieval facilities. They demonstrate no new concepts and can easily be simulated using the attribute list feature of IPL. These methods might be of some use in conjunction with a large parallel processing device. This is true, of course, of any scheme using coordinate indexing for retrieval.

All the memory organizations mentioned above have the following characteristics:

1.    Dynamic allocation of storage via available space lists or the equivalent.  This characteristic relates most directly to the updating problem in information retrieval.  Storage schemes such as assignment of location sequentially by order of arrival make initial storage and later additions particularly simple and economical but retrieval and deletion become quite difficult.  Dynamic allocation of storage requires more processing time but yields simplified retrieval, deletion and addition methods.  A new item is analyzed and its place in the storage structure is determined by its content.  Then the next cell on the available space list is assigned to this item. Appropriate addresses are stored in the cell and in the storage structure to complete the linking of the new item into storage.  Retrieval is based on the same analysis process used in storage.  An item is found by searching the links which connect it to the items in the structure.

A characteristic of many information retrieval problems is the indeterminate number of items which must be grouped together in storage.  The open-endedness of the dynamic method of storage described here is of particular value in these problems.

2.    Storage and retrieval as a function of content-oriented tags or associations implicit in list structures.  Many problems which fall within the scope of information retrieval make reference to relations between items (e. g. between documents) as well as to the items themselves.  That is, it is often desirable to retrieve all documents which deal with a particular topic, or which make reference to another particular set of documents or authors.  Information of this type can be stored either explicitly in tags (the tag may be, in the extreme case, the entire document) or implicitly in a linked-list structure.  A list structure is often more desirable because it easily permits growth and contraction.

B.    Parallel Searching

Devices with parallel search capabilities have appeared in the areas of cryogenics, semiconductors, magnetics and printed circuit cards. The advantage of parallel search appears to be a great reduction in retrieval time. Instead of retrieving each word of an N word memory and comparing it with the query, it would be possible to compare all N words simultaneously. Then it would be necessary to interrogate N flag bits or to generate in sequence the addresses of all words matching the query. In a memory with M-bit words this could amount to a reduction of a factor of M in the search time. However, the difficulties and expense in building large (on the order of $10^6$ words) memories to provide this capability makes parallel search uneconomical at present. The particular difficulty is the apparent need for semi-conductor devices at each word location for use in identifying the matching words. An alternative solution requiring an additional sense line and recognizing circuitry for each bit position, rather than each word, is given in Lewin (10). The hardware requirements are greatly reduced but the recognition and readout algorithm is more complex.

Some important work in the parallel search area involves the use of thin film cryogenic circuitry for both the storage and the identification. The production method involves simultaneous depositing of the search logic and the memory elements. This is discussed in the Rajchman paper (11).

The common features of parallel search plans are:

1.  Storage and retrieval on the basis of fixed field tags (number of tags, positioning, etc., sometimes variable through programming). This feature would be most applicable to fixed field data such as personnel files or parts inventories. Variable length fields in a fixed sequence (e.g. a library catalogue card) or unformatted data (e.g. raw text from a document or abstract) could be searched for a specific item only with rather complex programming. At this point the time advantage of parallel search would probably be lost. [2]

2.  Basic retrieval in terms of a conjunction of tags (other logical functions are sometimes available through programming).

This characteristic refers to the mask facility of the query register. Under normal operation, most proposals provide for the setting of the match flag for a word in memory if a match occurs with all the unmasked positions of the query register. Thus the logical conjunction of matches on any number of bits in the register may be formed. However, other logical operations are often desired in the search of a tagged memory. In particular, the negation operation will often be required. No parallel search proposal makes specific mention of hardware capable of this action. Negation could be simulated by programming but, again, the amount of programming and the additional memory cycles would overwhelm any timesaving from the original parallel search capability.

C.  Parallel Processing

The third basic property of associative memories appears in the work of Holland, Davies, Lee, Paull, Estrin and Fuller. Parallel processing involves the simultaneous operation of a great many very simple computing modules each with limited access to some of the others. To date the suggested applications have been largely in the area of numerical analysis (e. g. simultaneous calculations at several points for use in Runge-Kutta solutions to differential equations). An exception is found in the Lee and Paull paper (8) where algorithms are given for elementary retrieval functions using a one-dimensional, parallel-operating string processor.

More complex activities could be very useful in information retrieval. For example, local reorganization of a large memory could take place in response to recent user behavior. Since each module would be programmed separately, reorganization of a small section of the data could occur without searching or activation in other parts of the cell structure. This reorganization might be the addition to memory of items which are implied (in some specified sense) by items already stored. In a man-machine system this activity would take place without operator instigation or intervention.

The addition of logic capabilities to a previously passive memory structure is the common attribute found in all three types of associative memory mentioned in this report.  Much of the interest within information retrieval circles about associative memories arises from what are thought to be analogues to human subconscious retrieval and parallel processing abilities.  These analogues see the human brain as a very large linked store capable of very fast parallel search.  They usually fail to see the need for imbedded logic to implement such a device or the possibility that conscious serial processes might have behind them subconscious parallel processes.

## Conclusion

Active memories and parallel processing complexes can overcome the rigidities of the presently proposed parallel search devices. They could also implement the various list processing algorithms now in use.  The cost of this flexibility appears to be prohibitive but the trend toward faster and cheaper machines should soon overcome that problem.  The emphasis on improved batch fabrication techniques and integrated circuits is pushing computer design toward large modular parallel processors.  Initially such systems will seem uneconomical.  This is why information retrieval techniques utilizing these capabilities must be developed and demonstrated now.

### Notes

1.   This report was supported, in part, under task II of Contract AF19(628)-2752.

2.   The major value of parallel search memories may lie in their use in a control, rather than storage, function.  In such a situation they would be of small size and used perhaps to handle formatted program instruction sequencing.

### References

(1)  Cheydleur, B. F.  "SHIEF: A Realizable Form of Associative Memory," American Documentation Vol. 14, No. 1, Jan. 1963, p. 56.
A poorly conceived system using a complicated linked structure.

(2)  Davies, P.  "A Superconductive Associative Memory, "

Spring Joint Computer Conf., May 1962, San Francisco, p. 79.
Description of parallel search memory properties and proposals for
use of cryotrons.

(3)  Davies, P.  "Design for an Associative Computer," Pacific Computer Conference, IEEE, March 1963.
Computer organization combining parallel processing with parallel
search memory and designed using cryotron logic.

(4)  Estrin, G. and Fuller, R.  "Algorithms for Content Addressable Memory Organizations," Pacific Computer Conf., IEEE,
March 1963.
Uses for a content addressable memory and estimations of instruction execution times for various hardware implementations.

(5)  Goldberg, J. and Green, M. W., "Large Files for Information Retrieval Based on Simultaneous Interrogation of All Items"
in Yovits  Symposium on Large Capacity Memory Techniques for
Computing Systems, May 1961.
Million word memories, superimposed coding and various physical
realizations.

(6)  Griffith, J.  "Techniques for Advanced Information Processing Systems," Session 9, First Congress on the Information System Sciences, November 1962, p. 59.
Table look-up procedures for parallel search.

(7)  Kiseda, J. R. et al.  "A Magnetic Associative Memory
System," IBM Journal, Vol. 5, No. 2, April 1961, p. 106.
Parallel search memory using magnetic cores.

(8)  Lee, C. Y. and Paull, M. C.  "A Content Addressable
Distributed Logic Memory with Applications to Information Retrieval," Proceedings of the IEEE, June 1963, pp. 924-932.
Elementary hardware and micro-programming for a (1-dimensional)
cell memory.  For pattern matching all cells are searched in parallel while each cell is searched serially for the desired string.

(9)  Lee, E. S.  "Semiconductor Circuits in Associative Memories," Pacific Computer Conference, IEEE, March 1963, p. 96.
Word-organized memory, tag and information bits in each word, de-

signed with semiconductors.

(10)  Lewin, M. H.  "Retrieval of Ordered Lists From a Content-Addressed Memory," RCA Review, June 1962, pp. 215-229.
A good solution for the problem of retrieving multiple responses to a parallel search of memory.  The algorithms, but not the hardware, are compared with other methods.

(11)  Newhouse, V. and Fruin, R.  "A Cryogenic Data Addressed Memory," Spring Joint Computer Conf., May 1962, San Francisco, p. 89.
Description and experimental results relating to the design of a small cryotron system.

(12)  Prywes, N. S. and Gray, H. J.  "The Multi-list System," Technical Report No. 1, Moore School of Electrical Engineering, Univ. of Pa., 1961.
Partial solution to some of the retrieval problems encountered with list organized memories.

(13)  Rajchman, J. A.  "Magnetic Memories: Capabilities and Limitations," Computer Design Vol. 2, No. 8, September 1963, p. 32.
A survey article including sections on content addressable and superconductive memories.

(14)  Rosin, R. F.  "An Organization of An Associative Cryogenic Computer," Spring Joint Computer Conf., May 1962, San Francisco, p. 203.
Criterion of design and uses for a cryogenic system.

(15)  Seeber, R.  "Cryogenic Associative Memory," National Conference of ACM, August 1960, Milwaukee.

## IV.   The Data Processing Subsystem

The basic function of this subsystem could be termed problem-solving.   This term is here used so as to include searching (as distinct from lookup and recall), inference (deductive and inductive), problem selection, formulation and analysis.   Yet, the problem-solving process, as suggested in the Newell-Simon approach, can be viewed as a combination of storage/lookup (Chapter III) and transformation processes.

The problem-solving process begins when an agent in the community of "processors" recognizes a need, even though it be vaguely felt, ill-defined.   Through interaction with the knowledge subsystem and the storage/recall subsystem, to which he is linked by various bonds, he begins to formulate a goal.   The process of goal formulation and reformulation is a continuing one; if it is efficient and stable, each successive goal formulation is such that actions based on it would be increasingly effective in reducing the original need.

A well-formulated goal is compared with available resources to determine if they are relevant.   This is essentially a search process.   If there is no match, the goal is reformulated, transformed into subgoals, and the search operation is done on these. This process of transforming goals into subgoals, subgoals into subsubgoals, etc. is at the heart of problem analysis.   If matching a goal with stored resources is search, decomposing a goal into subgoals for subsequent search corresponds to research.

While the processes of transformation are very important, we have not begun to study them, but have concentrated on some aspects of search processes only.   Search is essentially a probabil-

istic concept.   Only an exhaustive search can be completely systematic, and this is of little interest here.

Thus, what is of greatest interest are efficient search strategies and devices to aid in searching.   One such device is that of
grouping items in the universe of search that are sufficiently similar so that a search passing through an item in such a group would
also pass "close" to the other items in that group.   Hopefully, the
chances for selecting items thus encountered and leading to successful search trails are greater than in the absence of such grouping,
and the chances of pursuing spurious search trails are much reduced.

Some significant algorithms for determining such groupings,
treated as a search for certain types of subgraphs in a graph, are
described in paper IV A.   These results are of interest in their own
right and have numerous and diverse applications.   One such application, as another useful by-product of this research, is described
in paper IV B.   Again, the key role of graph-theoretic notions is
apparent as a foundation for theoretical work in information science.

These ideas have also led to plans for a number of experiments using the AMNIP system and the clustering algorithms.   Some,
involving a thesaurus and citations, have already been described.
An additional one is described in paper III B which, like part of II C,
describes a practical by-product of this kind of research in a different field.   This paper is best viewed as a companion piece to IV A,
but it does not directly fit in this chapter.

The work on a growing thesaurus also involves questions of
organizing memory (for efficient searching as well as for efficient
storage).   Some elementary results to indicate the problem and what
can be done are described in paper IV C.   This paper should be
considered in conjunction with II D and III B.   The reason for including it in this chapter is that processing includes continual reorganization of memory to increase the effectiveness of searching.

It will be recalled that the most evident and important concrete embodiment of the processing subsystem in an information
system is a community of people in their roles as problem-solvers,

decision-makers, generators and authenticators of new knowledge. A person in this community acts as an agent of the social process whereby a new document--regarded as a cultural artifact--is synthesized from a combination of existing documents, plus a great deal of knowledge and reflective power, plus the resultant of all the socio-cultural forces exerted on this agent through his interaction with colleagues. It is these influences which helped determine which small faction and which mix of the total literature in the libraries this particular agent will be exposed to during his lifetime; which aspects of knowledge will register with him and become thoroughly assimilated into his internal representation of the world. These forces may help shape his needs, hence influence his problem-selection and his entire problem-solving process.

What do we know about the types of contacts a given agent is influenced, stimulated by, and how these affect his function as an information processor? Among the little that is known is the fact that direct personal contact is still the best and most widely used source of information and references; also that people aggregate into cliques or "clusters." A given agent may belong to more than one cluster. Various methods to exhibit these aggregations among people--and how they change--have been and are being studied. One of these studies was aimed at engineering a system to exploit these (dynamic) aggregations among people for improving the flow of information among them and thus, indirectly, help improve their processing/problem-solving capability. Such a system, now under experimental test at the University of California, Berkeley, is described in papers IV D and IV E by Flood and Kochen.

It is by no means clear that such systems as DICO and SASIDS (see IV D and IV E) provide services which are of greater net value than existing means of communication. The idea of a dynamic system responsive to use and changes in the environment is inherent in AMNIPS and the MEMEX idea. How an information user should and would allocate his limited time for communication requires deeper analysis and experimentation, and the papers reported in this section indicate a direction for progress.

# IV A.  Graph Theoretic Techniques for the Organization of Linked Data [1]

C. T. Abraham

Abstract:  The problem of finding subgraphs of a given graph which are strongly connected within themselves and weakly connected to their complement subgraphs has been related to the problem of "clustering." Both directed and undirected graphs have been considered.  Several algorithms for tracing circuits and proper paths have been obtained and used to get clusters.  In addition, when there are two distinct sets and a relation defined between these two sets, this will result in a bipartite graph.  When the bipartite graph is undirected, it is possible to obtain partitions of the two sets such that those subsets of the two sets which are strongly connected can be identified.  Procedures for accomplishing this have been briefly described.

For our discussion, by linked data  we mean either of the following:  1)  A set of items or objects and a binary relation defined on this set.  2)  Several sets of items and a correspondence between the sets.  For every pair of objects in the same set or in different sets, the binary relation or correspondence may be expressed in terms of:  1)  a similarity measure,  2)  an association factor,  3) a connectivity index which is either zero or one, etc.  Whenever any measure of similarity or association is used, a suitable threshold value for this measure is chosen and the binary relation or correspondence between pairs of objects is judged as either significant if larger than the threshold value or insignificant otherwise.  Thus in an attempt to form "near synonyms" in the English language,

229

Sparck-Jones used a measure of "near synonymity" of words in context and "clustered" words that are "near synonyms." V. Clapp makes reference to the problem of removing inconsistencies from a set of index terms on which a partial ordering, such as index term A is more general than index term B, has been defined. In an AFOSR report, "Adaptive Man-Machine Concept-Processing" by M. Kochen, C. T. Abraham and E. Wong, the idea of regarding the memory as a graph whose vertices stand for names and whose edges represent predicates, has been indicated. All the information stored in the memory is in the form of certain simple sentences, called RLR sentences, where the R's represent names and the L's represent predicates. "V. Bush is the author of 'As We May Think'" is one such sentence. In the storing and retrieving of such information, it is pertinent to investigate ways of organizing such an information file which will result in optimum use of the file for retrieval of facts on file relevant to specific queries. There are scores of other situations in information processing and electronic circuit design where the optimum organization of linked data appears to be the central intellectual problem.

In our attempt to organize linked data, the main emphasis has been on obtaining theoretical insights and algorithms which effect the organization. The following factors were considered in the development of the techniques for organization:

1) Ability to deal with large amounts of data.

2) Ability to optimize the organization of the whole data by optimum organization of parts of it. (When the problem is viewed as a systems analysis problem, optimization of the system is implied by the optimization of subsystems. Equivalently, if the optimum organization of linked data is achieved in stages, then optimality of any particular stage guarantees optimality of succeeding stages.)

3) Dynamic reorganization capabilities when new data are added.

Our main concern in this paper is the formation of "clusters" or groups. But solutions to other problems such as hierarchical structuring, pathtracing, etc. are obtained as by-products.

According to an earlier statement, linked data consists of a set or sets of objects and a binary relation R defined among the objects which can be represented by a graph whose vertices correspond to the objects and whose edges correspond to the binary relation.   This graph could be an n-partite graph when there are n sets of objects and the relation is defined between the sets.

We shall first deal with graphs which have only a single set of vertices.   The binary relation defined on the set of vertices could be any one of the following types:

Type I      The binary relation R is reflexive, non-symmetric and non-transitive.

Type II     The binary relation R is reflexive, symmetric and non-transitive.

Type III    The binary relation R is reflexive, non-symmetric and transitive.

Type IV     The binary relation R is reflexive, symmetric and transitive.

Note:   When the transitivity condition holds, then the binary relation R induces a partial ordering on the set of vertices.

We shall use two properties of the graph C which are class properties of the set of vertices V(G) and the set of edges E(G) of the graph to define clusters.

In almost all interesting applications when clusters are required, a decomposition of the graph into disconnected subgraphs will not be possible.   However, we shall obtain certain partitions of the set of all vertices.   For each of the disjoint sets of vertices in the partition, the section graph, i.e. the subgraph whose vertices are the elements of the set and whose edges are the edges between the vertices of the set, will be used to define clusters.   Such a subgraph is called a leaf graph.   The partition of the vertex set is accomplished by using an equivalence relation defined on the set of vertices.   This equivalence relation is called cycle-connectivity of vertices.   Two vertices, A and B, are said to be cycle-connected if there is a path from A to B and a path from B to A.   For a

graph with a large number of vertices, the partition of the vertex
set into cycle-connected subsets of vertices can be accomplished
with ease as shown in an algorithm to be described later.   Having
accomplished this, we will examine the section graphs (which are
much smaller than the original graph).   We will obtain further decomposition of these section graphs into still smaller graphs with
greater strength of connectivity within each such smaller graph.
This is accomplished by considering a class property of the edges
of the original graph.   The algorithms for generating the partition
of the set of edges will involve more detailed computation than the
one for obtaining the vertex set partition.   But it will be shown that
in order to obtain the partition of the set of edges, we need consider only the edges associated with the vertices of each subset of the
partition of the vertex set of the graph obtained by cycle connectivity.   Thus the increase in the computation is offset by a reduction
in the size of the graph.   The subgraphs whose edges are sets of
the partition of the edges of the original graph are called lobe
graphs.   Each leaf graph is either a lobe graph or it is the union
of lobe graphs, any pair of the latter having at most one vertex in
common.   The algorithm for finding lobe graphs will trace all the
circuits, Euler paths and redundant paths in the leaf graphs.   By
supplying the oriented edges between leaf graphs, all the paths in
the original graph can be obtained.

   Before describing the algorithms for clustering, we shall list
some of the graph theoretic justifications for using leaf and lobe
graphs for obtaining clusters:

   1.   Each leaf graph is circuit closed and every pair of vertices
in the leaf belongs to a circuit. [A circuit is a closed unicursal path
and a cycle is a union of circuits.   A unicursal path in a graph is
an unoriented chain whose edges can be oriented and arranged in a
sequence so that the terminal point of one edge is the initial point
of the next edge.]

   2.   Leaf decomposition partitions the vertices into mutually exclusive classes.

3.  Any leaf graph is a union of its lobe graphs; i.e. if L is a leaf and if $L_i^*$'s are the lobes in the leaf, then

$$G(L) = \sum_i G(L_i^*)$$

4.  Any two lobes can have at most one vertex in common.

5.  The leaf graphs of an undirected connected graph are at most two edge connected. [In a connected undirected graph, any subgraph is two edge connected if the removal of these two edges makes the original graph unconnected.]

6.  When the local degrees of an undirected graph are not even, the graph permits a leaf decomposition. [The local degree or the degree of a vertex V in a graph is the number of edges of the graph incident to V.]

7.  Every edge of attachment of a leaf graph is a separating edge. [If an edge E with vertices $V_1$ and $V_2$ is such that one of its vertices $V_1$ lies in a set S and the other vertex lies in $\{V(G)-S\}$, then E is called an edge of attachment of G(S). An edge E = $(V_1, V_2)$ is called a separating edge in a graph G if in the graph H obtained from G by the removal of E the vertices $V_1$ and $V_2$ are disconnected. ]

8.  Every vertex of attachment of a lobe graph is a separating vertex. [A vertex V in a section graph G(H) of a graph G(S) is called a vertex of attachment of G(H) to G(S-H) if the removal of V makes G(H) and G(S-H) unconnected. In a graph G, a vertex V is said to be a separating vertex if there is a proper nonvoid subgraph G' with V as its only vertex of attachment. ]

9.  Between any two leaves, in an undirected graph, there can at most be one separating edge.

10.  In a directed graph, the separating edges between any two leaves have the same orientation.

11.  For any directed graph with a leaf decomposition, one can obtain a new graph whose nodes are the leaves of the initial graph. Such a graph is called the leaf composition graph of the original graph. The leaf composition graph is a tree.

At this point we wish to elaborate on the concept of a cluster in a directed graph. We have mentioned earlier that we are interested in a subset of the set of vertices which have high interconnectivity among themselves and very little interconnectivity with vertices outside. The leaf graphs have the unique and very desirable property that connections back and forth can only exist within a leaf. The connections between two leaves are always in one direction. Thus if interconnectivity is required of the members of a cluster, then the most logical procedure is to obtain partitions of vertices and within each subset of vertices look for strong interconnectivity. The idea of finding the lobe graphs of each leaf graph accomplishes this latter purpose. The strong edge circuit connectivity will not guarantee complete subgraphs, but it will impose a slightly stronger condition on the interconnectivity of vertices within a leaf. The lobe graphs have at most one vertex in common for any pair. The decision to assign a common vertex of two lobes to one of them can be made by any one of several reasonable criteria. These criteria could be based solely on the actual number of edges from the vertex to and from the lobes or it could be based on the ratio of the number of vertices in a lobe to which the separating vertex is connected to the number of vertices in that lobe which are connected to it.

An interesting application of clustering for a directed graph is the formation of a hierarchy among index terms. The directed edge between any two index terms is established by which one of the index terms is more general. Thus the index term "abstracting" is more general than the term "automatic abstracting." This fact may be indicated by giving this edge a direction so that "abstracting" is the initial point and "automatic abstracting" the terminal point. If two index terms are synonyms, then between the corresponding vertices in the graph two edges with opposing orientation will be established. In this manner it is possible to form a directed graph among a set of index terms. It is evident that the leaf graphs of this graph correspond to terms in the same level in the hierarchical ordering of the index terms whereas the connections between leaves will help us to induce a partial ordering among the leaves.

We shall now discuss algorithms for obtaining leaf and lobe graphs
for a given graph.

## Case 1.  Directed Graph

We shall mean by directed graph a graph where the binary relation defined on the vertices is reflexive, non-symmetric and non-
transitive.  We shall give algorithms for the leaf and lobe decomposition of a directed graph.  The first algorithm for leaf decomposition uses the reachability matrix of a directed graph.  This technique had been first used by Harary for the reduction of a matrix
to find its eigenvalues.  Associated with a directed graph is a matrix which is an incidence matrix of vertices.  Thus if a directed
graph G has n vertices, then associated with it is an n x n matrix,
N.  If $1, 2, 3, \ldots n$, are used to denote the n vertices of the graph,
the $(i, j)^{th}$ entry of the matrix N is 1 if there is an edge from i to
j and 0 otherwise.  If I denotes the identity matrix, then $M^{n-1} =$
$(N+I)^{n-1}$ where matrix multiplication is Boolean, is called the reachability matrix of the directed graph G.  We shall denote the reachability matrix by P.  Obviously the $(i, j)^{th}$ entry of P is 1 if there is
a directed path from i to j in the graph.  In many directed graphs,
the longest path from any vertex i to any vertex j may have less
than $(n-1)$ edges.  Thus if the longest path is of length r, then evidently $P = M^{r+1}$.  If the $(i, j)^{th}$ entry as well as the $(j, i)^{th}$ entry
are both 1, then there is a directed path from i to j as well as
from j to i.  We shall call any two vertices i and j possessing this
property strongly cycle connected.  It will be shown that all vertices
which are strongly cycle connected with a given vertex form a leaf
of the vertex.

Theorem A     Strong cycle connectivity is an equivalence relation on
the set of vertices of a graph and the partition of the vertex set on
the basis of this equivalence relation gives all the leaves of the
graph.

Proof:  The existence of directed paths from vertex i to vertex j
and back means that vertices i and j belong to a cycle.  This cycle
is either the trivial cycle θ or it is the union of circuits.  [A triv-

ial cycle is the union of circuits each of which involves only two vertices.] Since a leaf graph is circuit closed and since in the cycle every pair of circuits has at least one vertex in common, all the circuits which form the cycle lie in a leaf.   Thus all the vertices which are strongly cycle connected with a specified vertex lie in the leaf of that vertex.

Corollary 1   If in a directed connected graph, each vertex is as often an initial vertex as it is a terminal vertex for edges of the graph, then the graph always has a proper lobe composition.

Proof:   The proof follows from the following theorem from König's "Theorie der Graphen," page 29.

Theorem        In a directed finite graph G, if each vertex is as often an initial point as it is a terminal point of edges of G, then there exists a system $S = \{C_1, C_2, \ldots, C_v\}$ of mutually (edge) disjoint circuits such that each edge belongs to one and only one member of the system S.

By our definition, a lobe consists of all those edges which are strongly circuit edge connected, i.e. there are circuits $C_1, C_2, \ldots,$ $C_k$ which have at least one edge in common such that any two edges $E_1$ and $E_2$ of a lobe belong to the sequence of circuits $C_1, C_2, \ldots,$ $C_k$ with $E_1$ in $C_1$ and $E_2$ in $C_k$.   By König's theorem, the lobes of the graph G are simply the circuits $C_1, C_2, \ldots, C_v$ of the system S. When $v = 1$, the entire graph forms a lobe.

An Algorithm for Obtaining the Leaves of a Directed Graph

Step 1        For $i=1, 2, \ldots, n$, count the number of 1's in the $i^{th}$ row and $i^{th}$ column of the matrix N.   If for each i, these numbers are equal, lobe composition is possible.

Step 2        Form the matrix $M = N+I$ and obtain the reachability matrix P by Boolean multiplication.   A method for computing the reachability matrix which is suitable for a digital computer program is as follows:

Method of Computing P.

Scan the first row of M and find all the 1's and their corresponding column numbers.   Let $j_1, j_2, \ldots$ denote the column numbers of 1's

appearing in row 1. Then OR the $j_1, j_2$ rows of M into row 1.
Then go to the rows 2, 3, ..., n and repeat the same operations.
When once all the rows have been scanned this way, we get a new
matrix, say $M_1$. Repeat the same operations with rows of $M_1$ resulting in a matrix $M_2$. Continue until two matrices $M_i$ and $M_{i+1}$
are identical. Then $M_i$ = P is the reachability matrix.

Each $M_r (r \le i)$ can be stored in the same location as $M_{r-1}$
and the time to form P is proportional to $n^2$ where n is the order
of N. (A computer program has been written for this.)

Step 3    Start with the first row of P. If the $i^{th}$ element of the
first row and the $i^{th}$ element of the first column are both 1, then
the vertices 1 and i are strongly cycle connected. Repeating this
operation with all the nonzero elements of the first row, we obtain
all the vertices of the graph which are in the same leaf with vertex
1. If the first vertex gives no leaf, then go the second row and repeat the same operations and so on. In this manner, we obtain
leaf decomposition of the graph.

Step 4    Delete the rows and columns of P corresponding to the vertices of the leaf graph in Step 3.

<u>Lobes of a Leaf</u>    Each leaf L in a directed graph G has a section
graph G(L) called the leaf graph of L. In order to obtain the lobe
graphs $G(L_i^*)$ in G(L) we have to obtain all the circuits in G(L).
Theorem C, below, and Corollary 1 will help to decide whether a
leaf graph has lobe composition or not.

Theorem B    The connecting edges of a leaf graph in a directed
graph are all either directed toward the leaf or away from the leaf.
Proof: If there are two edges with opposite orientation between two
leaves, then there will be a circuit with vertices of both leaves.
This is not possible.

Theorem C    A leaf graph with a Hamilton circuit is a lobe graph.
[A Hamilton circuit of a graph is a circuit which passes through
each vertex in the graph only once.]
Proof: A leaf graph is circuit closed and any two vertices lie in a
circuit. Since the leaf graph has a Hamilton circuit, every vertex

lies in one circuit.    Thus every edge connecting any two vertices
should be in a circuit involving at least two strongly circuit edge
connected edges.    This means the edge itself is strongly circuit
edge connected.

## Algorithm 1 for Tracing Circuits in a Leaf Graph

Take any vertex $i_0$ in the leaf graph G(L) as a starting point
and pass through the various directed edges from $i_0$ to their end
points.    Each edge $(i_0, i_1)$ is marked once as one leaves $i_0$ and it is
marked at $i_1$ as an entering edge.    If $i_1$ is not the initial point of
any edge, then $(i_0, i_1)$ is marked as closed.    The vertices which are
two edges away from $i_0$ (edges are taken with proper orientation)
are obtained by taking each of the edges from $i_0$ and marking it
again.    At $i_1$, each of the edges with $i_1$ as the initial point are
travelled and marked.    These may lead back to $i_0$ in which case we
list the pair of vertices $i_0$ and $i_1$ as belonging to a circuit of two
edges.    After this we return to $i_0$ by the entering edge.    If $(i_0, i_1)$
is closed, then we start on another edge $(i_0, i_2)$ and from $i_2$ go to
other vertices which can be reached by traversing edges which have
$i_2$ as the initial point.    As before, we will get circuits of length 2.
If $(i_0, i_2)$ is not closed, then return to $i_0$ and mark the edge $(i_0, i_2)$
again.    Continue this operation until all the edges at $i_0$ with $i_0$ as
an initial point have been marked twice.    Repeat the same operations until all vertices k edges away from $i_0$ have been reached and
we have returned to $i_0$ so that each edge with $i_0$ as initial point
have been marked k times and all edges with $i_1$ as initial point have
been marked (k-1) times and so on.    To go to the vertices which
are (k+1) edges away from $i_0$, we move successively to all vertices
like $i_1$, at one edge away and go to all vertices k edges away from
them.    Each time we obtain the circuits and keep record on them.
In addition, we list all the vertices visited.    [This algorithm applies
to any directed graph.    If all the vertices of a given graph have not
been traversed in this manner, i.e. if the given graph has no Hamilton circuits, then we consider any vertex which has been left out
by the paths obtained previously and continue with that as the initial

vertex. Continue this until all vertices have been used. ]

The above algorithm will also give all the trees in the graph which are required in problems other than clustering.

## Algorithm II for Tracing Circuits in a Directed Graph

We have seen that in order to obtain the lobes in a leaf graph of a directed graph, we have to obtain all the circuits in the leaf graph. In addition, we are quite often interested in obtaining all the Euler paths which exist in a directed graph. Here we shall propose an algorithm for tracing circuits and Euler paths in a directed graph. The algorithm involves matrix multiplication which can be performed on a digital computer. The size of the graph to which this algorithm can be applied will be predicated on computer memory limitations.

## Algorithm

Associated with a directed graph with n vertices is an n x n matrix whose $(i, j)^{th}$ entry is either a prime number or zero depending on whether there is a directed edge from vertex i to vertex j or not. Thus, each edge in the graph is identified by a distinct prime number. A path connecting any two vertices i and j will be made up of edges whose prime numbers are, say, $p_1, p_2, \ldots p_t$. Thus, the path from i to j is represented by $p_1, p_2, p_3 \ldots p_t$. The absence of a path will be designated by 0. When there are several paths connecting any two vertices i and j such as $p_{i_1}, p_{i_2}, p_{i_3}, \ldots p_{i_t}$; $p_{j_1}$, $p_{j_2}, \ldots p_{j_s}$ etc. this will be indicated by the sum $p_{i_1}, p_{i_2}, \ldots p_{i_t} + \ldots$

Numerical summation is not allowed except the addition of 0, i. e. $p_i, p_2 \cdots p_t + 0 = p_1, p_2 \cdots p_t$. If M is the n x n matrix whose elements are prime numbers or zero, depending on whether there are edges between pairs of vertices or not, then $M^r$ will give all the paths of length r in the directed graph.

The construction of circuits and Euler paths is simplified by the fact that a path which is neither a Euler path nor a circuit always contains a circuit whose length is smaller than that of the path.

The matrix $N^r$ which contains entries of only Euler paths and circuits is obtained from the matrices $M^s$ ($s \le r$) when the following additional rules are applied. After $M^s$ ($s \ge 1$) is computed add all diagonal entries to a "circuit list." Repeat this operation after the construction of every $M^s$, so that when $M^r$ is formed, the list will contain all circuits of length $(r-1)$ or less. In addition, delete from $M^s$ any off-diagonal element which is divisible by any of the terms appearing in the updated "circuit list." In $N^r$, thus obtained, all the diagonal terms would represent circuits of length $r$ and off-diagonal elements would represent Euler paths of length $r$. Before computing $M^{s+1}$, all diagonal terms as well as all off-diagonal elements divisible by circuits are deleted from $M^s$.

### Illustrative Example

Let

$$M = \begin{array}{c} \\ 1 \\ 2 \\ 3 \end{array} \begin{array}{ccc} 1 & 2 & 3 \\ \begin{pmatrix} 0 & 0 & 3 \\ 2 & 7 & 11 \\ 0 & 5 & 0 \end{pmatrix} \end{array}$$

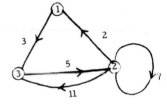

$$M^2 = \begin{array}{c} \\ 1 \\ 2 \\ 3 \end{array} \begin{array}{ccc} 1 & 2 & 3 \\ \begin{pmatrix} 0 & 3 \cdot 5 & 0 \\ 2 \cdot 7 & 7^2 + 5 \cdot 11 & 2 \cdot 3 + 7 \cdot 11 \\ 2 \cdot 5 & 2 \cdot 7 & 5 \cdot 11 \end{pmatrix} \end{array}$$

$$M^3 = \begin{array}{c} \\ 1 \\ 2 \\ 3 \end{array} \begin{array}{ccc} 1 & 2 & 3 \\ \begin{pmatrix} 2 \cdot 3 \cdot 5 & 2 \cdot 3 \cdot 7 & 3 \cdot 5 \cdot 11 \\ 2 \cdot 7^2 + 2 \cdot 5 \cdot 11 & 2 \cdot 3 \cdot 5 + 7^3 + 5 \cdot 7 \cdot 11 + 2 \cdot 7 \cdot 11 & 2 \cdot 3 \cdot 7 + 7^2 \cdot 11 + 2 \cdot 5^2 \cdot 11 \\ 2 \cdot 5 \cdot 7 & 5 \cdot 7^2 + 5^2 \cdot 11 & 2 \cdot 3 \cdot 5 + 5 \cdot 7 \cdot 11 \end{pmatrix} \end{array}$$

$$N = \begin{array}{c} \\ 1 \\ 2 \\ 3 \end{array} \begin{array}{ccc} 1 & 2 & 3 \\ \begin{pmatrix} 0 & 0 & 3 \\ 2 & \textcircled{7} & 11 \\ 0 & 5 & 0 \end{pmatrix} \end{array}$$

$$
N^2 = \begin{array}{c} \\ 1 \\ 2 \\ 3 \end{array}
\begin{array}{ccc} 1 & 2 & 3 \\ \left( \begin{array}{ccc} 0 & 0 & 3 \\ 2 & 0 & 11 \\ 0 & 5 & 0 \end{array} \right) \end{array}
\times
\begin{array}{c} \\ 1 \\ 2 \\ 3 \end{array}
\begin{array}{ccc} 1 & 2 & 3 \\ \left( \begin{array}{ccc} 0 & 0 & 3 \\ 2 & 0 & 11 \\ 0 & 5 & 0 \end{array} \right) \end{array}
=
\begin{array}{c} \\ 1 \\ 2 \\ 3 \end{array}
\begin{array}{ccc} 1 & 2 & 3 \\ \left( \begin{array}{ccc} 0 & 3\cdot 5 & 0 \\ 0 & \boxed{5\cdot 11} & 2\cdot 3 \\ 2\cdot 5 & 0 & 2^3\cdot 7 \end{array} \right) \end{array}
$$

$$
N^3 = \begin{array}{c} \\ 1 \\ 2 \\ 3 \end{array}
\begin{array}{ccc} 1 & 2 & 3 \\ \left( \begin{array}{ccc} 0 & 0 & 3 \\ 2 & 0 & 11 \\ 0 & 5 & 0 \end{array} \right) \end{array}
\times
\begin{array}{c} \\ 1 \\ 2 \\ 3 \end{array}
\begin{array}{ccc} & 2 & 3 \\ \left( \begin{array}{ccc} 0 & 3\cdot 5 & 0 \\ 0 & 0 & 2\cdot 3 \\ 2.5 & 0 & 0 \end{array} \right) \end{array}
=
\begin{array}{c} \\ 1 \\ 2 \\ 3 \end{array}
\begin{array}{ccc} 1 & 2 & 3 \\ \left( \begin{array}{ccc} \boxed{2\cdot 3\cdot 5} & 0 & 0 \\ \boxed{2\cdot 5\cdot 11} & \boxed{2\cdot 3\cdot 5} & 0 \\ 0 & 0 & \boxed{2\cdot 3\cdot 5} \end{array} \right) \end{array}
$$

The circled numbers correspond to members of the circuit list. The term in square corresponds to the off diagonal elements which are multiples of circuits.

Thus the vertex sets forming circuits including loops are: {2} {2,3} {1,2,3}. The vertex sets forming Euler paths are the remaining off-diagonal elements in N, $N^2$ and $N^3$.

## Illustrative Example for Clustering

We shall apply the algorithm for obtaining leaves to find clusters by considering a directed graph with 10 vertices. The N matrix of the graph is as follows:

|       | 1 | 2 | 3 | 4 | 5 | 6 | 7 | 8 | 9 | 10 |
|-------|---|---|---|---|---|---|---|---|---|----|
| 1     | 0 | 0 | 0 | 1 | 1 | 0 | 0 | 0 | 0 | 1  |
| 2     | 1 | 0 | 1 | 1 | 0 | 0 | 0 | 0 | 0 | 0  |
| 3     | 0 | 0 | 0 | 1 | 0 | 0 | 1 | 1 | 0 | 0  |
| 4     | 1 | 0 | 0 | 0 | 1 | 0 | 1 | 0 | 0 | 0  |
| 5     | 1 | 0 | 0 | 1 | 0 | 1 | 1 | 0 | 0 | 0  |
| 6     | 0 | 0 | 0 | 0 | 0 | 0 | 1 | 0 | 0 | 0  |
| 7     | 0 | 0 | 0 | 0 | 0 | 0 | 0 | 0 | 0 | 1  |
| 8     | 0 | 0 | 0 | 1 | 0 | 0 | 0 | 0 | 1 | 0  |
| 9     | 0 | 1 | 0 | 0 | 0 | 0 | 0 | 0 | 0 | 1  |
| 10    | 0 | 0 | 0 | 0 | 0 | 1 | 0 | 0 | 0 | 0  |

$N =$ (applies to the matrix above)

|       | 1 | 2 | 3 | 4 | 5 | 6 | 7 | 8 | 9 | 10 |
|-------|---|---|---|---|---|---|---|---|---|----|
| 1     | 1 | 0 | 0 | 1 | 1 | 1 | 1 | 0 | 0 | 1  |
| 2     | 1 | 1 | 1 | 1 | 1 | 1 | 1 | 1 | 1 | 1  |
| 3     | 1 | 1 | 1 | 1 | 1 | 1 | 1 | 1 | 1 | 1  |
| 4     | 1 | 0 | 0 | 1 | 1 | 1 | 1 | 0 | 0 | 1  |
| 5     | 1 | 0 | 0 | 1 | 1 | 1 | 1 | 0 | 0 | 1  |
| 6     | 0 | 0 | 0 | 0 | 0 | 1 | 1 | 0 | 0 | 1  |
| 7     | 0 | 0 | 0 | 0 | 0 | 1 | 1 | 0 | 0 | 1  |
| 8     | 1 | 1 | 1 | 1 | 1 | 1 | 1 | 1 | 1 | 1  |
| 9     | 1 | 1 | 1 | 1 | 1 | 1 | 1 | 1 | 1 | 1  |
| 10    | 0 | 0 | 0 | 0 | 0 | 1 | 1 | 0 | 0 | 1  |

$P =$ (applies to the matrix above)

The Connectivity matrix after obtaining leaves is as follows:

|        | 1 | 4 | 5 | 2 | 3 | 8 | 9 | 6 | 7 | 10 |
|--------|---|---|---|---|---|---|---|---|---|----|
| 1      | 1 | 1 | 1 | 0 | 0 | 0 | 0 | 0 | 0 | 1  |
| 4      | 1 | 1 | 1 | 0 | 0 | 0 | 0 | 0 | 1 | 0  |
| 5      | 1 | 1 | 1 | 0 | 0 | 0 | 0 | 1 | 1 | 0  |
| 2      | 1 | 1 | 0 | 0 | 0 | 0 | 0 | 0 | 0 | 0  |
| 3      | 0 | 1 | 0 | 1 | 1 | 0 | 0 | 0 | 1 | 0  |
| 8      | 0 | 0 | 0 | 1 | 1 | 1 | 0 | 0 | 0 | 1  |
| 9      | 0 | 0 | 0 | 0 | 1 | 1 | 1 | 0 | 0 | 1  |
| 6      | 0 | 0 | 1 | 0 | 0 | 1 | 1 | 1 | 1 | 1  |
| 7      | 0 | 0 | 0 | 0 | 0 | 0 | 0 | 1 | 1 | 1  |
| 10     | 0 | 0 | 0 | 0 | 0 | 0 | 0 | 1 | 1 | 1  |

$N_1 =$

Now using the algorithm for obtaining the leaves of a directed graph, we obtain the leaves as $\{1, 4, 5\}$  $\{6, 7, 10\}$  and  $\{2, 3, 8, 9\}$. We also notice that these leaves have Hamilton circuits in each one of them so that they are also lobes.

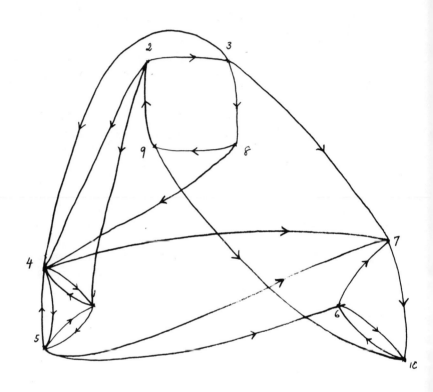

The directed graph with ten vertices

Figure 1

## Case 2.   Directed Graph with Transitivity

By a directed graph with transitivity we mean a graph in which if there is an edge from vertex $V_1$ to vertex $V_2$ and another edge from vertex $V_2$ to vertex $V_3$, then there is an implied edge from vertex $V_1$ to vertex $V_3$. If G is any directed graph, then the graph obtained by adding all the implied edges which are not already in G is called the transitive closure graph $G_t$. The following theorem from the theory of graphs asserts that the leaf graph of $G_t$ can be obtained from the leaf graph of G.

Theorem     If G is a directed graph with single edges, then the transitive closure graph $G_t$ is a graph whose leaf graphs are the directed complete subgraphs $G^{(d)}(L_i)$ defined on the leaf sets $L_i$ of G while the leaf composition graph $G'_t$ of $G_t$ is a partial order isomorphic to the transitive closure of the leaf composition graph $G'$ of G.

In fact, the leaf graphs of $G_t$ are simply the complete directed subgraphs obtained from the leaf graphs of G. In addition, if we consider the leaf composition graphs of G and $G_t$, then the leaf composition graph $G'_t$ of $G_t$ is a partial order isomorphic to the leaf composition graph $G'$ of G. By a leaf composition graph we mean the homomorphic mapping of a graph which maps each leaf onto a vertex. Such graphs are circuit free.

This means that we could operate with the graph G rather than with $G_t$ and then complete the leaf graphs of G to obtain clusters in $G_t$. Since the leaf graphs of $G_t$ are complete, they have no lobe decomposition. If G is irreducible, so is $G_t$.

## A directed graph with a very large number of vertices

The algorithm for obtaining all the leaves of a directed graph involves Boolean multiplication of 0-1 matrices. Depending on the number of 1's and the order of this matrix, there is a limit to the maximum size of the directed graph that can be analyzed without increasing the computation time exponentially. We shall discuss a procedure which will avoid this possibility. Let m be the maximum cardinality of the set of vertices which permits economic computation

time.    Then for a directed graph with n vertices where n >> m,
the leaf graphs can be obtained by the following procedure:

Step 1    Take any m vertices and form the leaf graphs of the section graph involving these vertices.

Step 2    Take all the vertices in the graph to which or from which
there are directed edges with the m vertices chosen in Step 1.    Let
$m_1$ denote the total number of such vertices.

Step 3 (a)    If $m_1$ is less than m, then add $(m-m_1)$ vertices to the
initial $m_1$ vertices from among the vertices to which or from which
there are edges with the $m_1$ vertices.    If the number $(m-m_1)$ is
less than the number of vertices connected by edges to the $m_1$ vertices, then rank these vertices in decreasing order of the number
of edges to or from the $m_1$ vertices and choose the first $(m-m_1)$ of
the ranked vertices.

Step 3 (b)    If $m_1$ is greater than m, then rank the $m_1$ vertices in
descending order on the basis of the number of connecting edges to
or from the m vertices in Step 1 and choose the first m of them.

Step 4    Keep a list of all the one-edge connections between the m
vertices in Step 1 and the m vertices in Step 3 and with directions.

Step 5    Apply the algorithm for obtaining leaf graphs to the m vertices chosen in Steps 2 and 3.    List all the vertices in each leaf.

Step 6    Repeat Steps 2, 3, 4 with the vertex set in Step 4 as the
initial set of Step 1.

Step 7    Iterate all the previous steps till every vertex has been
used.

Step 8    Suppose the iteration had been performed r times.    Then
for each of the r section graphs, there is a leaf composition graph.
Now combine these r homomorphic graphs and supply the connecting
links obtained in Step 4 after deletion of duplicate connections between leaf graphs.

Step 9    Obtain the circuits that exist in the combined composition
graph of Step 8 by using the algorithm for tracing circuits and proper paths described earlier.

Step 10    From the circuit list, form classes of vertices which are cycle connected.    Then the leaf graphs thus obtained are the leaf graphs of the original graph.

Step 11    List all the vertices of the original graph with n vertices which are in the leaf graphs of Step 10.    These are the leaf graphs we want.

If there is high interconnectivity in the original data, the number of vertices in the leaves may be too many to economically apply the circuit tracing algorithm to obtain the lobes.    But for highly interconnected data, there is very little hope to gain optimum organization which may have any practical use.

Step 12    Within each lobe graph, all the proper paths and circuits can be traced by one of our algorithms.

Dynamic reorganization when new vertices are added to a directed graph:

The solution to this problem is essentially outlined in the preceding section.    To start with, there is a directed graph with $n_o$ vertices, the leaf composition graph of which can be obtained by applying our algorithms.    When new vertices are added we simply apply the procedure outlined with the exclusion of Steps 6 and 7 if the number of added vertices is less than m.    Otherwise everything goes through and the leaf and lobe graphs as well as circuits and Euler paths can be obtained.

Case 3.    Undirected Graph

If in a connected undirected graph the local degree of each vertex is even, then the graph is irreducible (no leaf decomposition). If, however, the graph is unconnected, then the leaf decomposition will give the connected components.    The lobe decomposition is still possible for an undirected connected graph.    In fact, there should be at least two lobes which are one vertex attached.    As in the directed graph, here also we can obtain clusters.    The algorithms for tracing paths in a leaf graph for a directed graph can be used with some modifications to trace the paths in an undirected graph.

Case 4.    Undirected Graph with Transitivity

By an undirected graph with transitivity we mean a graph in

which if there is an edge between vertex $V_1$ and vertex $V_2$ and an edge between vertex $V_2$ and vertex $V_3$, then there is an implied edge between vertices $V_1$ and $V_3$. It has been indicated earlier that an undirected connected graph has no leaf decomposition. If, however, the graph is unconnected, then transitive closure will give the completion of the connected components, i. e. every connected component in the original graph G will give rise to a completion graph in the transitive closure.

## Algorithm for decomposition of an undirected bipartite graph

The algorithm we discuss here is based on the method of selection of distinct representatives due to Marshall Hall and the transversal algorithm due to Dulmage and Mendelsohn. We shall not attempt to give detailed description of the work of these authors.

Let $S_1$ and $S_2$ be two sets of m and n elements, respectively. The correspondence between $S_1$ and $S_2$ is represented as a matrix $A = || a_{ij} ||$ where $a_{ij}$ is either zero or one and $1 \leq i \leq m$, $1 \leq j \leq n$. It is assumed that $a_{ij} \neq a_{ji}$ for all i and j. We are interested in row and column permutations of the matrix A such that the permuted matrix will have most of the non-zero elements along and near the main diagonal.

Procedure:

Step 1    If the matrix A of the undirected bipartite graph satisfies the condition that there exists at least one i for which $a_{ii} = 0$, then from the expansion of determinant $| A |$, pick a non-zero term $a_{i_1 j_1}, a_{i_2 j_2} \ldots a_{i_m j_n}$ where $i_1 \neq i_2 \neq i_3 \neq \ldots \neq i_m$ and $j_1 \neq j_2 \neq j_3 \neq$

$\ldots = j_n$. Such a term is called a transversal. The choice of the transversal can be accomplished by algorithms due to Marshall Hall, Dulmage and Mendelsohn, etc. Dulmage and Mendelsohn have devised algorithms which are easily adaptable to computer programming and these algorithms are quite satisfactory when m and n are large numbers.

Step 2    Permute the rows and columns of the matrix $|| a_{ij} ||$ such

that the elements of the transversal $a_{i_1 j_1}$, $a_{i_2 j_2} \ldots a_{i_m j_n}$ appear
along the main diagonal of the matrix $\|a_{ij}\|$. This can be accomplished by an algorithm for partitioning of a bipartite graph due to Dumage and Mendelsohn. If A! is the matrix which is obtained by row and column permutations, then A' is in the decomposed form.

As an illustrative example, we consider a 7 x 7 matrix A associated with a bipartite graph.

$$
A = \begin{pmatrix}
0 & 0 & 1 & 0 & 0 & 1 & 0 \\
0 & 1 & 0 & 0 & 0 & 0 & 1 \\
1 & 0 & 0 & 1 & 0 & 0 & 0 \\
0 & 1 & 0 & 1 & 0 & 0 & 0 \\
1 & 0 & 1 & 0 & 1 & 1 & 0 \\
1 & 1 & 0 & 0 & 1 & 0 & 0 \\
1 & 0 & 0 & 1 & 0 & 0 & 0
\end{pmatrix}
$$

Using Steps 1 and 2 of the algorithm, the row and column permutations are found as follows:

Row permutations: (1 7 2 5 6 4 3)

Column permutations: (1) (2 3 6 7 5 4)

The elementary matrices associated with these permutations can be now obtained. Thus, if P is the row permutation matrix and Q is the column permutation matrix, then:

$$
A' = PAQ =
\begin{array}{c c}
 & \begin{array}{ccccccc} 1 & 4 & 2 & 5 & 7 & 3 & 6 \end{array} \\
\begin{array}{c} 3 \\ 7 \\ 4 \\ 6 \\ 2 \\ 5 \\ 1 \end{array} &
\begin{pmatrix}
1 & 1 & 0 & 0 & 0 & 0 & 0 \\
1 & 1 & 0 & 0 & 0 & 0 & 0 \\
0 & 1 & 1 & 0 & 0 & 0 & 0 \\
1 & 0 & 1 & 1 & 0 & 0 & 0 \\
0 & 0 & 1 & 0 & 1 & 0 & 0 \\
1 & 0 & 0 & 1 & 0 & 1 & 1 \\
1 & 0 & 0 & 0 & 0 & 1 & 1
\end{pmatrix}
\end{array}
$$

The numbers outside the matrix give the identification of the members of the two sets. A' gives the desired decomposition.

The algorithm for the decomposition of a bipartite graph can be extended to n-partite graph with some modifications.

Note: For a directed graph with n vertices which has a Hamilton circuit, the algorithm for decomposition of a bipartite graph can be used provided there exists a non-zero transversal. Thus in the illustrative example above, the square matrix A could very well be the matrix of a directed graph. In that case, it is easy to verify that the directed graph has a Hamilton circuit 1 6 5 3 4 2 7 1 where the vertices are taken in the order in which they are written. Then A' is the decomposition. However, to form clusters, the matrix A' is of very limited use.

Computer programs

The algorithm for finding the leaf graphs in a directed graph has been programmed for the IBM 7094. The program will handle a graph with 2,000 vertices and takes four minutes on the average.

Another program for tracing paths and circuits is being written now.

Program Description--Algorithm for Clustering by Matrix Operations

Method

The program has been written in FAP for the IBM 7094. The program runs under FORTRAN Monitor System, Version II. For the Boolean matrix multiplication, the method suggested by J. J. Baker in Communications of the ACM is used in general. Only those elements of the matrix are represented in core which are one's. Each element corresponds to one word; the row is stored in the decrement, the column in the address.

Usage

Data can be read from: a. Tape, written as a one record file where row is followed by column, in binary. b. Cards.

1.    Input on tape

    a.    Use tape unit 4 to read the input tape.

    b.    The number of items to be read in from the tape is en-

tered on a card with the following format (1X, 110).    Two of
these cards are needed.

    c.   Put sense switch 1 down.

2.    Input on cards

    a.   Each element of the matrix which is a one is entered on
a separate data card.    The first word on the card is the row
designation, the second refers to the column.    Use the following format (1X, 2120).    The last card is followed by a blank
card.

    b.   Put Scratch tape on tape unit 5.

Limitations

    The upper limit of the dimension of the initial matrix is 2000
by 2000.    The maximum number of one's in the final matrix after
multiplication may not exceed 24, 000.

Output

The output contains:

    a.   The new index arrangement after clustering.    The matrix
is divided into areas; boxes along the diagonal containing the
strongly connected components and the horizontal and vertical areas
occupying the space between the box and the boundary of the matrix.
The lowest box on the diagonal does not, of course, have any attached area.

    b.   The number of occupants at one side of the boxes which
is to be multiplied by itself to get the total dimensions of the particular box.

    c.   The ratio of 1-s to 0-s inside the boxes.

    d.   The position of the 0-s inside the boxes.

    e.   The ratio of 0-s to 1-s inside both the vertical and hori-
zontal areas.

    f.   The position of 1's inside both the vertical and horizontal
areas.

### Note

[1]This work was supported in part under Contract AF19(628)-2752.

IV B.  An Application of Clustering Techniques to Minimizing
the Number of Interconnections in Electrical Assemblies[1]

C. T. Abraham

Abstract:  In many instances, the packaging of electronic systems
is accomplished by grouping electrical subassemblies into larger assemblies.   Subject to certain constraints concerning the number and
type of subassemblies for each assembly, the number of distinct
signals per assembly, the power consumption by each assembly,
etc., a grouping of subassemblies into assemblies has to be performed so as to minimize the number of interconnections between
assemblies.   A procedure for doing this is outlined in this paper
along with an application to a problem discussed by Lawler.

In many instances, the packaging of electronic systems is accomplished by grouping electrical subassemblies into larger assemblies.   The basic subassemblies are logic gates which are manufac-
tured as printed circuits or modules.   Assemblies may be made up
of subassembly boards placed on "mother boards," or of miniature
wafers arranged in stacks.   Often the type of wiring between assemblies is quite different from that within each assembly.

We shall address ourselves to the problem of grouping subassemblies into assemblies in such a way that the smallest possible
number of electrical connections is required between the resulting
assemblies.   In addition, we shall require that within each assembly
the interconnectivity is maximum.

This grouping or clustering problem had been investigated by
Lawler (2), Muroga (3) and others.   We shall start with Lawler's
formulation of this problem.   Each subassembly has terminals that

are to be electrically common with certain specified terminals in the other assemblies. Each set of electrically common terminals is identified with an electrical signal. Conversely, each subassembly is characterized by the electrical signals that are identified with its terminals. If we use the integers $1, 2, 3, \ldots$ to designate the signals, then each subassembly $A_i$ is characterized by the signal set

$$S_i = \{ i_1, i_2, \ldots, i_{n(S_i)} \}$$

where $n(S_i)$ = the number of elements in the set $S_i$.

   $= n(A_i)$ = the number of signals identified with

   subassembly $A_i$.

When subassemblies are grouped together into assemblies, each assembly is characterized by the signals identified with its member subassemblies. Thus, if assembly B is composed of subassemblies $A_p$, $A_q$, $A_r$, then the signal set of B is

$$S_p \cup S_q \cup S_r = \{ j_1, j_2, \ldots j_{n(B)} \}$$

where $n(B)$ = number of distinct signals identified with one or more of the subassemblies contained in assembly B.

There might be constraints on the formation of assemblies such as

1) The number of member subassemblies $\leq k$

2) The total power consumption of member subassemblies $\leq m$.

3) The number of member subassemblies of a certain type $\leq L$.

4) The number of distinct signals identified with member subassemblies $\leq n$.

Problem Statement

Here we shall formulate the problem of grouping together subassemblies into assemblies so as to minimize the number of interconnections between assemblies subject to the following constraints.

1) The number of distinct signals identified with member sub-

assemblies in any assembly should not exceed k.

2) The number of signals shared by member subassemblies in an assembly is maximum.

3) The number of member subassemblies of a certain type in an assembly is less than a specified number.

Method of solution. We shall propose a method of solution to to this problem using graph theory. Thus our first objective is to reduce the problem formulation to one in terms of a graph. There is a choice of using either a directed graph or an undirected graph. A directed graph is much more amenable to the type of analysis we wish to perform for reasons explained in reference (1). Now, each subassembly corresponds to a node of the graph and for any two subassemblies $A_i$ and $A_j$, the decision as to whether there should be a directed edge from Ai to Aj or from $A_j$ to $A_i$ has to be made on the basis of some measure of connectivity of the subassemblies. Let $n_i$, $n_j$ and $n_{ij}$ be the number of signals in $A_i$, the number of signals in $A_j$ and the number of common signals in $A_i$ and $A_j$, respectively. Then the ratio of signals of $A_i$ which are common with $A_j$ to the total number of signals in $A_j$ will be used as a measure of the fractional overlap of $A_j$ with $A_i$. The fractional overlap of $A_i$ with $A_j$ will be measured by $\dfrac{n_{ij}}{n_i}$ . The decision to establish an edge from $A_i$ to $A_j$ or from $A_j$ to $A_i$ will be based on this fractional overlap by the choice of a suitable threshold value whose choice will be detailed in the sequel.

Let $A_1$, $A_2$, ... $A_n$ be the n subassemblies and $S_1$, $S_2$, ... $S_n$ be their corresponding signal sets.

Let $n_i$, $n_j$ and $n_{ij}$ be as defined earlier.

Step 1

For every pair $A_i$ and $A_j$ (i, j = 1, ... n) form $n_i + n_j - n_{ij}$ $= d_{ij}$. Evidently $d_{ij}$ is the total number of distinct signals in either $A_i$ of $A_j$.

Define: $a_{ij} = \begin{cases} 1 \text{ if } n_i + n_j - n_{ij} \leq k. \\ 0 \text{ otherwise} \end{cases}$

This information is stored in the form of subassembly x subassembly incidence matrix A

$$A = || a_{ij} ||$$

Step 2

For every $(A_i, A_j)$ pair for which $a_{ij} = 1$

Compute
$$r_{ij} = \frac{n_{ij}}{n_i}$$

$$r_{ji} = \frac{n_{ij}}{n_j}$$

Form the matrix $R = ( r_{ij} )$

The main diagonal entries $r_{ij}$ are all 1's and for any $r_{ij}$, $i \neq j$, $0 \leq r_{ij} \leq 1$.

Step 3

For all $r_{ij}$, $i \neq j$, obtain an ordering. Let $\alpha$ be the greatest lower bound and $\beta$ the least upper bound of this ordering.

i. e.,
$$\alpha = \min. \{ r_{ij}; i \neq j \}$$
$$\beta = \max. \{ r_{ij}; i \neq j \}$$

Step 3a

Map the matrix R onto R', $R' = (r'_{ij})$, where
$$r'_{ij} = \begin{cases} 1 & \text{if } r_{ij} \geq \alpha \\ 0 & \text{if } r_{ij} = \alpha \end{cases}$$

Step 4

Form $R' - I = M$, where I is the n x n identity matrix.

Count the number of 1's in the first row and first column.

Let this be denoted by $b_1$.

Continue this operation to determine $b_2$, $b_3$, ... $b_n$.

The numbers $b_1$, $b_2$, ... $b_n$ are the local degrees of the vertices (subassemblies).

Step 5

If all the $b_i$'s are even, then go to Step 3 and repeat Step 3 by mapping R onto $R^{(2)} = (r''_{ij})$ where

$$r^{(2)}_{ij} = \begin{cases} 1 & \text{if } r_{ij} > a_1 \\ 0 & \text{if } r_{ij} \leq \alpha_1 \end{cases}$$

where $\alpha_1 = \min \{r_{ij}; \ i \neq j \text{ and } r_{ij} > a\}$

Then repeat Step 4 with $R^{(2)}$.

Repeat this operation till the $R^{(i)}$ is obtained where the $b_i$'s are not all even.

Let the matrix $R^{(i)} - I$ be denoted by N.

Step 6

Apply the algorithm for obtaining a leaf decomposition of the directed graph D(N) associated with the matrix N. This algorithm consists in obtaining the reachability matrix of D(N) and finding the cycle connected vertices [see Ref. (1)].

A computer program for this has been written. The program handles a matrix of order 2000 with less than 5 per cent nonzero entries in the matrix. There also exists a program which will take a matrix of order 500 without any restrictions on the number of non-zero entries in the matrix.

Let $L_1, L_2, \ldots L_r$ be the leaves of D(N).

Step 7

Take each leaf $L_i$ and find the number of distinct signals shared by the subassemblies in it. If in any leaf all the subassemblies satisfy the constraint that the total number of distinct signals is less than k, then that leaf is an assembly. If now we want to impose the constraint that the total number of subassemblies in an assembly is less than, say, t, then we form the lobes of the leaf which satisfy the signal constraint. Then each lobe could form an assembly. If there are lobes which contain more subassemblies than allowed, then for each subassembly in a lobe, find its connectivity number and rank the subassemblies in the decreasing order of connectivity constant. Pick the highest ranking subassembly and find all the nodes which are one edge away (distance 1) from it. Among these subassemblies at a unit distance, obtain a ranking on the basis of increasing number of connections to subassemblies outside this set and the initial subassembly. The highest ranking subassembly is added to the initial subassembly. Then repeat the same operation with the two subassemblies as the initial ones. Continue adding new subassemblies until the required number is obtained. Thus if $A_{i_1}$, $A_{i_2}$, ... $A_{i_t}$ are the chosen subassemblies form the lobe which will be on an assembly, then form the sets

$A_{i_1}^{(1)}$, $A_{i_1}^{(2)}$, $A_{i_1}^{(3)}$ which are sets of subassemblies at distances

1, 2, 3,... from $A_{i_1}$. Similarly form $A_{i_2}^{(1)}$, $A_{i_2}^{(2)}$, $A_{i_2}^{(3)}$ ...

$$A_{i_3}^{(1)}, A_{i_3}^{(2)}, A_{i_3}^{(3)}, \ldots$$

$$A_{i_t}^{(1)}, A_{i_t}^{(2)}, A_{i_t}^{(3)}, \ldots$$

It is easy to show that in each of the above sequences, any subassembly in any set can at most be connected to subassemblies in the preceding and succeeding sets in its sequence.

Consider the set intersections

$$\bigcap_{j=1}^{t} A_{i_j}^{(1)}, \quad \bigcap_{j=1}^{t} A_{i_j}^{(2)}, \ldots, \quad \bigcap_{j=1}^{t} A_{i_j}^{(x_0)}$$

If $x_0$ is the maximum value of $x$ for which

$$\bigcap_{j=1}^{t} A_{i_j}^{(x)}$$

is nonempty, then pick a subassembly

in $\bigcap_{j=1}^{t} A_{i_j}^{(x_0)}$ and continue the operation used for picking

$$A_{i_1}, \ A_{i_2}, \ldots, \ A_{i_t}$$

Let this new set be $A'_{i_1}, \ A'_{i_2}, \ \ldots, \ A'_{i_t}$

Again we form

$$\bigwedge_{j=1}^{t} A'^{(1)}_{i_j}, \ \bigwedge_{j=1}^{t} A'^{(2)}_{i_j}, \ldots, \ \bigwedge_{j=1}^{t} A'^{(x)}_{i_j}$$

Then we form

$$\bigwedge_{j=1}^{t} A'^{(x)}_{i_j} \ \bigwedge \ \bigwedge_{j=1}^{t} A'^{(x)}_{i_j} \ = \ U^{(x)}, \ \text{say}.$$

Then find $x_o$, the largest $x$ for which $U^{(x_o)}$ is nonempty and contains a subassembly not among $Ai_j$'s and $Ai'_j$'s. Then pick a

subassembly from $U^{(x_o)}$ satisfying the above requirement and proceed as before. This operation is continued until all the subassemblies in the lobe have been assigned to assemblies.

To obtain the sets $A^{(x)}_{i_j}$, a program can be written using the

principle of progressive covering. [See Ref. (1).]

Step 8

If any leaf $L_1$ does not satisfy the condition that the number of distinct signals of all the subassemblies in the leaf should be less than k, then form its lobes and check out each lobe for number of distinct signals and repeat Step 7.

The above method of solution will give a good assignment with the constraints concerning the number of member subassemblies and the number of distinct signals in an assembly. The solution also maximizes the number of signals shared by subassemblies in an assembly.

Numerical Example

For purposes of illustration of our method we shall use the

same problem as Lawler (2).

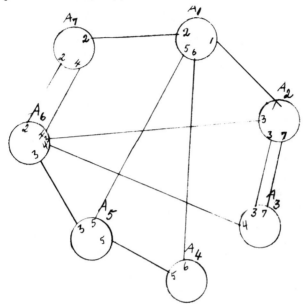

Suppose the subassemblies are to be combined into assemblies subject to the following restrictions:

1)    No more than 3 subassemblies may be members of an assembly.

2)    No more than 4 distinct signals may be identified with an assembly.

3)    The number of signals shared by subassemblies in the same assembly is maximum.

$$A_1 = \{1, 2, 5, 6\}$$
$$A_2 = \{1, 3, 7\}$$
$$A_3 = \{3, 4, 7\}$$
$$A_4 = \{5, 6\}$$
$$A_5 = \{3, 5\}$$
$$A_6 = \{2, 3, 4\}$$
$$A_7 = \{2, 4\}$$

Step 1

$$A_1 U A_2 = \{1, \ 2, \ 3, \ 5, \ 6, \ 7\}$$
$$A_1 U A_3 = \{1, \ 2, \ 3, \ 4, \ 5, \ 6, \ 7\}$$
$$A_1 U A_4 = \{1, \ 2, \ 5, \ 6\}$$
$$A_1 U A_5 = \{1, \ 2, \ 3, \ 5, \ 6\}$$
$$A_1 U A_6 = \{1, \ 2, \ 3, \ 4, \ 5, \ 6\}$$
$$A_1 U A_7 = \{1, \ 2, \ 4, \ 5, \ 6\}$$
$$A_2 U A_3 = \{1, \ 3, \ 4, \ 7\}$$
$$A_2 U A_4 = \{1, \ 3, \ 5, \ 6, \ 7\}$$
$$A_2 U A_5 = \{1, \ 2, \ 3, \ 4, \ 7\}$$
$$A_2 U A_6 = \{1, \ 2, \ 3, \ 4, \ 7\}$$
$$A_2 U A_7 = \{1, \ 2, \ 3, \ 4, \ 7\}$$
$$A_3 U A_4 = \{3, \ 4, \ 5, \ 6, \ 7\}$$
$$A_3 U A_5 = \{3, \ 4, \ 5, \ 7\}$$
$$A_3 U A_6 = \{2, \ 3, \ 4, \ 7\}$$
$$A_3 U A_7 = \{2, \ 3, \ 4, \ 7\}$$
$$A_4 U A_5 = \{3, \ 5, \ 6\}$$
$$A_4 U A_6 = \{2, \ 3, \ 4, \ 5, \ 6\}$$
$$A_4 U A_7 = \{2, \ 3, \ 4, \ 5\}$$
$$A_5 U A_6 = \{2, \ 3, \ 4, \ 5\}$$
$$A_5 U A_7 = \{2, \ 3, \ 4, \ 5\}$$
$$A_6 U A_7 = \{2, \ 3, \ 4\}$$

Step 2

|       | $A_1$ | $A_2$ | $A_3$ | $A_4$ | $A_5$ | $A_6$ | $A_7$ |
|-------|-------|-------|-------|-------|-------|-------|-------|
| $A_1$ | 1 | 0 | 0 | 1 | 0 | 0 | 0 |
| $A_2$ | 0 | 1 | 1 | 0 | 1 | 0 | 0 |
| $A_3$ | 0 | 1 | 1 | 0 | 1 | 1 | 1 |
| $A_4$ | 1 | 0 | 0 | 1 | 1 | 0 | 1 |
| $A_5$ | 0 | 1 | 1 | 1 | 1 | 0 | 1 |
| $A_6$ | 0 | 0 | 1 | 0 | 0 | 1 | 1 |
| $A_7$ | 0 | 0 | 1 | 1 | 1 | 1 | 1 |

$A =$ (matrix above, with row labels $A_1$ through $A_7$)

|     | $A_1$ | $A_2$ | $A_3$ | $A_4$ | $A_5$ | $A_6$ | $A_7$ |
|-----|-------|-------|-------|-------|-------|-------|-------|
| $A_1$ | 1 | 0 | 0 | $\frac{1}{2}$ | 0 | 0 | 0 |
| $A_2$ | 0 | 1 | $\frac{2}{3}$ | 0 | $\frac{1}{3}$ | 0 | 0 |
| $A_3$ | 0 | $\frac{1}{3}$ | 1 | 0 | $\frac{1}{3}$ | $\frac{2}{3}$ | $\frac{1}{3}$ |
| $A_4$ | 1 | 0 | 0 | 1 | $\frac{1}{2}$ | 0 | 0 |
| $A_5$ | 0 | $\frac{1}{2}$ | $\frac{1}{2}$ | $\frac{1}{2}$ | 1 | 0 | 0 |
| $A_6$ | 0 | 0 | $\frac{2}{3}$ | 0 | 0 | 1 | $\frac{2}{3}$ |
| $A_7$ | 0 | 0 | $\frac{1}{2}$ | 0 | 0 | $\frac{1}{2}$ | 1 |

$R =$ (the matrix above)

Step 3

$$\frac{1}{3} < \frac{1}{2} < \frac{2}{3} < 1$$

$$\alpha = \frac{1}{3}$$

$$\beta = \frac{2}{3}$$

Step 3a

|     | $A_1$ | $A_2$ | $A_3$ | $A_4$ | $A_5$ | $A_6$ | $A_7$ |
|-----|-------|-------|-------|-------|-------|-------|-------|
| $A_1$ | 1 | 0 | 0 | 1 | 0 | 0 | 0 |
| $A_2$ | 0 | 1 | 1 | 0 | 0 | 0 | 0 |
| $A_3$ | 0 | 0 | 1 | 0 | 0 | 1 | 0 |
| $A_4$ | 1 | 0 | 0 | 1 | 1 | 0 | 0 |
| $A_5$ | 0 | 1 | 1 | 1 | 1 | 0 | 0 |
| $A_6$ | 0 | 0 | 1 | 0 | 0 | 1 | 1 |
| $A_7$ | 0 | 0 | 1 | 0 | 0 | 1 | 1 |

$R' =$ (the matrix above)

Step 4

|  | $A_1$ | $A_2$ | $A_3$ | $A_4$ | $A_5$ | $A_6$ | $A_7$ |
|---|---|---|---|---|---|---|---|
| $A_1$ | 0 | 0 | 0 | 1 | 0 | 0 | 0 |
| $A_2$ | 0 | 0 | 1 | 0 | 0 | 0 | 0 |
| $A_3$ | 0 | 0 | 0 | 0 | 0 | 1 | 0 |
| $A_4$ | 1 | 0 | 0 | 0 | 1 | 0 | 0 |
| $A_5$ | 0 | 1 | 1 | 1 | 0 | 0 | 0 |
| $A_6$ | 0 | 0 | 1 | 0 | 0 | 0 | 1 |
| $A_7$ | 0 | 0 | 1 | 0 | 0 | 1 | 0 |

$R' - I = M$

The local degree is not even for each vertex
$$b_1 = 2 \qquad b_2 = 3 \qquad b_3 = 5 \qquad b_4 = 4 \qquad b_5 = 4 \qquad b_6 = 3 \qquad b_7 = 2$$

Step 5     is unnecessary

Step 6

The algorithm for leaf graph.  We will perform this by actually looking at the graph since it is small

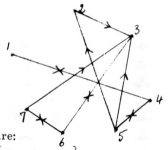

The leaves are:
$$L_1 = \left\{ A_1,\ A_4,\ A_5 \right\} \qquad\qquad L_2 = \left\{ A_3,\ A_6,\ A_7 \right\}$$

Step 7

We examine leaf $L_1$ for satisfying the constraint that the number of distinct signals $\leq 4$.
$$(A_1 U A_4) = \left\{ 1,\ 2,\ 5,\ 6 \right\}$$

So $A_1$ and $A_4$ form a subassembly.

We now examine leaf $L_2$.  This leaf is    $A_3$, $A_6$, $A_7$.

Now we examine $A_3 U A_6 U A_7$ for the number of distinct signals. They are   2, 3, 4, 7.   Thus,    $A_3$, $A_6$, $A_7$   forms an assembly.

Now we are left with $A_2$ and $A_5$.  We combine them into an assembly.   Thus, by leaf and lobe decomposition the optimum assignment is    $A_1$, $A_4$      $A_2$, $A_5$      $A_3$, $A_6$, $A_7$ .  This is optimum provided we are maximizing the interconnections between subassemblies in each assembly.

Assembly interconnections for optimal solution are:

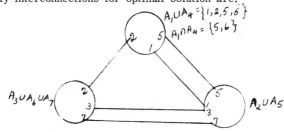

$A_3 U A_6 U A_7$ = 2,  3,  4,  7        $A_2 U A_5$ = 1,  3,  5,  7

$A_3 A_6$ = 3,  4     $A_3 A_7$ = 4        $A_2 A_5$ = 3

$A_6 A_7$ = 2,  4     $A_6 A_7 A_3$ = 4

The problem treated here has an important analogy with certain problems in information science.   We could, for example, treat a word as characterized by a set of words associated with it in various ways, analogous to the characteristic signal set for a subassembly.   We then wish to group words so that words within a word group--corresponding to assemblies--have maximum interconnectivity, and interconnections among different word groups are minimized. The method of solving these analogous problems are the same.

Note

1.   This work was supported, in part, under Contract AF19-

(628)-2752.


References

(1)  Abraham, C. T.  "Graph Theoretic Techniques for the Organization of Linked Data, "  pp. 229-251 of this volume; IV A.

(2)  Lawler, E. L.  "Electrical Assemblies with Minimum Number of Interconnections, " IRE transactions on Electronic Computers, Feb. 1962, pp. 86-88.

(3)  Muroga, S.  "Interim Report on the Study of Minimization of the Number of Interconnections, " April, 1963.  IBM Internal Report.

IV C.  A Note on Minimizing Search and Storage in a
Thesaurus Network by Structural Reorganization of the Net [1]

Phyllis Reisner

In "Constructing an Adaptive Thesaurus by Man-Machine Interaction" (1) a thesaurus system for Information Retrieval was described.  The thesaurus system, based on M. Kochen's "Adaptive
Man-Machine Non-Numerical Concept Processing System" (AMNIPS)
(2), consists of a machine-stored thesaurus and index which are to
be queried via a keyboard console.  The thesaurus itself is to be
"grown" by users of the system in the course of their own searches for
information and is expected to improve through use.  The computer, in
this man-machine system, is to serve as a linguistic data gatherer, recording the rare, but hopefully not unique, linguistic "acts" of individual, isolated users in a kind of "group memory" to benefit future ones.

In this man-machine thesaurus, efficient storage procedures
must be devised to balance the "man" costs (of search time and
frustration) against the "machine" costs (of access time, storage
redundancy and computing time).  In general, in any application of
the AMNIP system, problems of efficient memory organization to
obtain such balancing become critical, for the structuring and restructuring of the memory must be a continuing process built into
the system.  Since the users of the system "train" its memory by
entering information directly into the memory without benefit of human pre-editing, the burden of such pre-editing must fall on the
computer.  Thus programs to prevent duplication and/or contradic-
tion in the data entered must be incorporated into the system, as
must programs to evolve an increasingly more structured system

265

system from the initial haphazard, growed-like-Topsy one.    A first attempt at specifying such restructuring, or "adaptive reorganization," for the thesaurus is described below.

However, before discussing the restructuring, let us briefly review the mechanics of using the man-machine thesaurus.    To use the system, a querist enters index terms through the console and receives a display of alternative index terms to use in reformulation of his search query.    A man-machine conversation can ensue, in which the user enters a query word, receives a display of words associated with the query word, and repeats the procedure, selecting new words from the display at each step.    This "path-tracing" procedure, extremely difficult in purely manual systems, is less difficult but probably still tedious in man-machine systems.    We can organize the memory to decrease the number of such iterations, and consequent user time and frustration, by increasing the amount of redundant storage.    If, on the other hand, we wish to decrease storage, then we must pay for it by increasing the number of steps in the query reformulation "conversation."    We wish to balance these two factors by restructuring the memory.

The graph analogy of the AMNIP system provides a useful tool for specifying this restructuring.    The memory of this system is conceived as a graph in which, for the thesaurus application, the nodes correspond to index terms and the links to relations, such as subsumption, between the terms.    Data are entered into the system in the form of statements which correspond to a pair of nodes together with the link between them (e. g. [word a] [is synonymous with] [word b]).    The restructuring of the memory is to be accomplished by addition and/or deletion of such node-link-node statements.

In this note, we are concerned only with words connected by one type of "link," morphological variation (e. g., abstractor, abstracting--words with the same stem and different inflectional and/ or derivational endings).    The relation, "having the same stem as," is symmetric, reflexive, and transitive, partitioning the thesaurus graph of words linked by this relation into mutually exclusive subgraphs. In terms of expected functioning within the system, these subgraphs are complete graphs, every "node" potentially connected to every other. (Practically, this means a user interested in any node in a subgraph should have displayed to him all other nodes in the subgraph.) We wish to reorganize these subgraphs of morphologically related words to balance user search time with machine storage space.

Let:  $n$  = total number of nodes in a connected subgraph N

$s$  = total number of links in N

$p_{ij}$  = distance between nodes i and j, where $p_{ij}$ is taken as the number of links in the shortest path between i and j

$d$  = diameter of N (distance between the furthest nodes in N) = $p_{ij}$ (max.)

Clearly, search time can be interpreted in terms of $p_{ij}$ (at each node in a path a machine look up of time t seconds is required and a man-machine conversation takes place). Storage can be interpreted in terms of s (to each s corresponds a statement [node-link-node] or partial statement of average length b bits). We desire a configuration such that, for a given n, both s (storage) and d (maximum search time) are minimal.

The minimum number of links $s_{min}$, required for a graph with n nodes to be connected is:

$$s_{min} = n - 1$$

the maximum number, $s_{max}$, beyond which we have duplication-- more than one link between any pair of nodes, is:

$$s_{max} = \frac{n(n-1)}{2}$$

The minimum diameter of a graph with n nodes is

$$d_{min} = 1$$

the maximum,

$$d_{max} = n-1$$

The maximum storage, however, corresponds to the minimum

search,

$$s_{max} = \frac{n(n-1)}{2}$$

$$d_{min} = 1$$

and conversely, the minimum storage corresponds to the maximum search

$$s_{min} = n-1$$

$$d_{max} = n-1$$

We can not, therefore, simultaneously minimize search and storage. However, we can reorganize the graph so that, by increasing search slightly from its minimum possible value $d = 1$ to $d = 2$, we can simultaneously achieve the minimum possible storage: $s = n-1$.

To do this, we simply pick any node to use as a "central" node. We "erase" all links in the graph and replace them with links that "tie" each node only to the central node. It is thus possible to proceed from any node to any other, by traversing the central node, in a 2-step path. In this configuration, the so-called "star" graph,

$$d = 2$$

$$s = n-1$$

It is clear that, here, the combined cost of storage (s) and maximum search (d) is minimized.

To demonstrate the difference in search and storage, several possible configurations of a subgraph, N, of related thesaurus terms are illustrated in Figure 1 (for $n = 5$). In a., the chain graph configuration, storage is minimized at the expense of search, in b., the complete graph, the reverse occurs. In c., we see the preferred star-graph configuration which balances the two (search and storage).

This reorganization balances the man "costs" of searching (number of iterations) with the machine "costs" of storage. It also balances machine costs of number of access "probes" with storage. If, for example, the average length of node-link-node statement, (s),

a.   Chain Graph                              b.   Complete Graph

p = 4                                              p = 1
s = 4                                              s = 10

c.   Star Graph

p = 2
s = 4

Figure 1.   Three Subgraph Configurations for n = 5

|                | Machine Search Time = p X 30 ms | Storage = s X 100 bits |
| -------------- | ------------------------------- | ---------------------- |
| Chain Graph    | 120 ms                          | 400 bits               |
| Complete Graph | 30 ms                           | 1000 bits              |
| Star Graph     | 60 ms                           | 400 bits               |

Figure 2.   Search Time vs. Storage

is 100 bits, and the average access time is 30 ms, then we would have the search time and storage requirements shown in Figure 2. The star graph is clearly preferable.

In this paper, therefore, we have:

1)   Raised the problem of balancing storage against search (number of iterations) in a man-machine system of the type discussed and suggested "adaptive restructuring" of the memory to achieve such balance.

2)   Used a graph analogy as a tool for specifying such restructuring.

3)   Specified a star-graph configuration as the optimal structure in one particular instance.

### Note

1.   This work was partly supported under Contract AF19(628)-2752.

### References

(1)   Reisner, P., "Constructing an Adaptive Thesaurus by Man-Machine Interaction," the Final Report on Contract AF19(626)-10, Quarterly Report Number 9 (in preparation). A more recent version of this paper appears in Appendix IV E.

(2)   Kochen, M., Abraham, C., and Wong, E., "Adaptive Man-Machine Concept-Processing," Final Report on Contract AF19-(604)-8446, June 1962.

IV D.  Some  Bibliographic  and  Sociological  Devices  to  Improve
Maintenance  of  Current  Awareness  About  Literature [1]

M.  M.  Flood

and

M.  Kochen

Abstract:  Systems  that  distribute  notices  of  information  automatically  to  a  group  of  participants,  to  maintain  current  awareness,  seek
to  maximize  relevancy  relative  to  irrelevancy.    A  new  type  of  automatic  information  dissemination  system  (DICO),  based  upon  similarities  of  interests  of  subgroups  of  participants,  has  been  found  promising  in  limited  tests  and  the  continued  development  of  such  systems
is  in  progress.    One  such  system  (SASIDS)  is  an  adaptive  network
for  dissemination  that  adjusts  automatically  to  take  account  of  feedback  from  recipients.

The  distribution  of  abstract  journals,  indexes,  tables  of  contents  and  similar  regular  notices  of  new  publications  has  long  been
a  major  device  to  help  maintain  current  awareness.    The  widespread
sale  of  Current  Contents  bears  witness  to  the  effectiveness  of  a  device  as  simple  as  periodical  publication  of  the  tables  of  contents  of
a  few  major  journals.

Nonetheless,  the  judgment  that  hit  and  acceptance-rate  of  such
devices  both  need  to  be  increased  led  to  experiments  with  the  selective  dissemination  of  information  (4).    Since  a  user  will  generally
allot  a  fixed  fraction  of  his  time  for  maintaining  current  awareness,
his  use  of  such  systems  depends  on  shifting  his  reliance  from  his
current  mix  of  information  sources  to  a  new  mix.    He  can  either
increase  the  range  of  titles  and  abstracts  from  which  to  make  read-

271

ing choices, up to his limit, or rely on the system to make some of the choices for him.

A reader could rely on journals to which he subscribes, to make the choices for him, because he has found them to have made good choices for him in the past. This is the most common procedure.

Second, he could rely on someone who selects journals for his general area of interest, as done by Current Contents, who then presents tables of contents as the choice for him.

Third, he could rely on his own ability to describe his reading interest by keywords, and rely on indexers with access to remote materials to flag notices of documents with index terms matching his interests, as in SDI (4).

Fourth, he could rely on the recommendations of colleagues, who in the past have recommended well (2).

Fifth, and finally, he could rely on a good newspaper which, in the past, has conveniently kept him informed with the latest needed news but few diversions. This latter, in the best tradition of successful practice in mass communication media--good journalistic style, reporting, editing, in the form of broadcasts or newsprint to be discarded--might well be the most effective method to maintain current awareness.

The great advantage of journalistic style is that it accommodates a mass audience; it is tailored more to individual interests than could be any response to profiles. It is personalized in that the reader can use as little or as much of the information as he needs, and special areas of interest are covered in special sections easily skipped or read.

The literature is fragmented according to subject matter. So are professional cliques. There should be one newspaper for each professional clique, with reporters covering all the "literature" that pertains to that clique. Actually, the reporters should report newsworthy facts prior to their documented publication. If it cannot cover all the "pertinent" work within a length a reader can handle, the clique is too big and ready for further fragmentation, or the collection of work is not suitably matched to the clique. If all work is not covered with all the newspapers, then there is too much work and the unreported work will probably go unnoticed. If there is too much duplication in reporting, then there are probably too many newspapers relative to the number of readers, hence demand and price should decrease.

The task of maintaining such a news service could be assisted by automation in various ways. The people who generate news are often not the best reporters of news. The job of a reporter can be aided by devices for informing him of possibly newsworthy sources. A centralized system like SDI informs all its users of newsworthy items. In a news service, all the potential readers--and news sources--send information to a center. The reporters and editors for such a news service compile, rank, and express newsworthy findings on their own initiative and modify their performance to continually satisfy the needs of their readership.

We have developed a system (DICO) for generating an automatic distribution of documents to members of a group of readers, based upon similarity of interests of several sub-groups or clusters of individuals. Each participant in a DICO system reports each document he finds relevant and interesting, omitting obviously standard publications, and the system then distributes an abstract of the document to all other members of each cluster that includes the reporting participant among its membership. DICO was tested in a limited experiment, and found promising as a new kind of automatic dissemination method (3).

To determine clusters in the DICO experiment, 30 scientists at the IBM Research Center who volunteered their participation, were each given a random sample of 200 recent articles, by title and author. To this, each respondent provided 200 interest/no interest responses. All pairs of rows on the resulting 30 x 200 incidence matrix were then compared to determine for which pairs the fraction of responses-in-common exceeded a given threshold. A cluster consisted of the largest collection of respondents such that each member had enough responses-in-common with every other member in that cluster.

Two overlapping six-member clusters were found.  Each member was asked to submit notices to unusual items in the literature that he deemed recommendable to his colleagues.  These notices were sent to all the colleagues in his cluster, as well as to the other respondents who served as a control group.  The acceptance-rate for notices received by a member of the same cluster as the originator was 4 - 9 times as high as that for notices received by members of the other cluster, or for notices to a random sampling of the literature.

An algorithm for finding clusters by non-exhaustive methods was developed and programmed, and is subject to continuing improvement.  Experiments wherein the DICO participants could pose technical questions, wherein interest profiles of members in a cluster were compared were also conducted.  A dummy individual, his interest profile being the composite of those of the members of the cluster he represented, was entered as participant in the SDI system.

We have devised another system (SASIDS) which is like DICO in that automatic distribution is made to participants upon the basis of reports of relevancy by other participants, but unlike DICO in that two transition probabilities for each pair of participants determine the probability that a document found relevant by one will therefore be sent to the other.  Thus, a stochastic dynamic distribution network determines the automatic dissemination, rather than clusters.  This Stochastic Adaptive Sequential Information Dissemination System (SASIDS) has not yet been tested adequately.  SASIDS has been used for a year on a very limited basis with thirty participants and seems promising.  A full-scale experiment is now in progress with SASIDS, under the Space Sciences Laboratory of the University of California at Berkeley, and will be reported upon later elsewhere.

Fortunately, SASIDS does not require any difficult or tedious calculations in actual operation.   There is a requirement for rather substantial storage capacity, if there are many participants, in order to keep a record of the current pair-wise transition probabilities.   A modest amount of computation is required to up-date the transition probabilities, after each information transaction for a pair, to implement the linear Markov learning model (1) that makes the network adapt using feedback reports from recipients regarding relevancies of documents they receive.

Natural evolution of automatic dissemination systems like SDI, DICO, and SASIDS may lead eventually to much more highly personalized newspapers.   Conceivably, the items selected each day for each particular individual subscriber would correspond to his current personal tastes and needs, as indicated by feedback from him.

### Note

1.   Preparation of this paper was partly supported under Contract AF19(628)-2752.

### References

(1)   Flood, Merrill M., "Stochastic Learning Theory Applied to Choice Experiments with Rats, Dogs, and Men," Behavioral Science, 7, No. 3, July 1962, pp. 289-314.

(2)   Dawe, Albert, R., "Scientific Information Coding and the Hibernation Information Exchange," American Scientist 49, No. 4, December 1961, pp. 344A-358A.

(3)   Kochen, M. and Wong, E., "Concerning the Possibility of a Cooperative Information Exchange," IBM Journal of Research and Development, 6, No. 1, April 1962, pp. 270-271.

(4)   Luhn, H. P., "Selective Dissemination of New Scientific Information with the Aid of Electronic Processing Equipment," American Documentation, 12, No. 2, April 1961, pp. 131-138.

## IV E.   A Stochastic Adaptive Sequential Information
Dissemination System—SASIDS

### Merrill M. Flood[1]

### Introduction

Manfred Kochen and Eugene Wong[2] have developed and experi-
mented with an information dissemination system, called DICO, that
was based upon cooperative exchange of notices about published articles.   The present paper describes in more detail than was done in
IV D, an alternative system, SASIDS, that retains the cooperative
exchange principle but differs substantially from DICO in other respects.

We start by describing DICO briefly.   We then make a comparison of an actual distribution made in a DICO experiment with
one that would have been made by an alternative system that em-
ploys the sequential principle used for SASIDS, and conclude that the
results favor such sequential procedures.   Finally, we propose one
specific sequential system that is also adaptive in the sense that the
later distributions made depend upon results of feedback responses
by each recipient during earlier stages.   This proposed system
(SASIDS) makes use of a stochastic coupling between pairs of participants to achieve adaptation while also continually improving the proportion of desirable distributions.

### DICO

Each participant in a DICO system first rates each of a group
of titles of articles according to the relevancy for his own work and
interests on a 4-point scale where a rating of 3 means extremely
relevant and a rating of 0 means quite irrelevant.   Alternatively,

the rating might be simply 1 or 0 for relevant or irrelevant, and the data used here are of this kind.

The resulting data can be presented in matrix form, where the entry $R_{ij}$ in the $i^{th}$ row and $j^{th}$ column is the relevancy rating given by Participant i to Abstract j; the rating matrix $R \equiv \| Rij \|$ therefore has elements that are 1 or 0. The elements $S_{ij}$ of the square symmetric matrix $S \equiv RR^T$ then denote the number of titles judged relevant by both Participant i and Participant j. Of course $S_{ii}$ denotes the number of titles judged relevant by Participant i.

DICO is characterized by the procedure for determining subsequent distributions, based upon a cluster analysis of the matrix S for an initial distribution in which each Participant rates each of several titles. Generally speaking, the cluster analysis divides the entire group of Participants into several subgroups, perhaps with overlapping between two or more subgroups, in such a way that a title judged relevant by any Participant within a subgroup is very apt to have also been judged relevant by all other Participants within that subgroup. So, DICO distributes each title contributed as relevant by a Participant to every other Participant sharing membership with the contributing Participant in any subgroup. The recipients no longer return relevancy ratings to DICO after the subgroups are determined, except as this was done in the DICO Experiment for purposes of evaluating the effectiveness of the system.

## DICO Experiment

The DICO Experiment reported by Kochen and Wong utilized 30 Subjects who rated more than 110 titles using a relevancy index taking values 0, 1, 2, and 3, where 0 indicated no relevancy and 1, 2, 3 indicated increasingly relevant items. For our present analysis we interpret ratings of 1, 2, and 3 simply as relevant and 0 as irrelevant, and we analyzed only the first 110 titles distributed and only Subjects 1, 2, 3, 4, 5, 6, 8, 9, 11, 13, and 14. Data for Subjects 7, 10 and 12 were omitted because they each rated less than 4 titles as relevant.

The value of S for these 11 Subjects and 110 titles follows.

$$S \equiv$$

| i\j | 1 | 2 | 3 | 4 | 5 | 6 | 8 | 9 | 11 | 13 | 14 |
|---|---|---|---|---|---|---|---|---|---|---|---|
| 1 | 4 | 1 | 1 | 1 | 1 | 2 | 1 | 1 | 1 | 1 | 1 |
| 2 | | 13 | 1 | 3 | 2 | 3 | 1 | 3 | 1 | 2 | 1 |
| 3 | | | 5 | 1 | 2 | 2 | 2 | 1 | 2 | 2 | 3 |
| 4 | | | | 4 | 2 | 2 | 1 | 2 | 3 | 3 | 0 |
| 5 | | | | | 5 | 1 | 2 | 2 | 1 | 3 | 1 |
| 6 | | | | | | 9 | 1 | 3 | 2 | 1 | 1 |
| 8 | | | | | | | 7 | 1 | 1 | 2 | 3 |
| 9 | | | | | | | | 5 | 2 | 2 | 0 |
| 11 | | | | | | | | | 6 | 3 | 0 |
| 13 | | | | | | | | | | 8 | 1 |
| 14 | | | | | | | | | | | 7 |

The next step is based on the notion that an item first introduced into the system as relevant to Subject i should be brought to the attention of Subject j if and only if past experience suggests that an "appreciable" portion of items relevant to i were also relevant to j.    For this trial run, we arbitrarily required that "appreciable" mean more than one-fourth.    This means that Subjects send on relevant titles as indicated in Table 1.

Table 1

| Initiator | Recipients |
|---|---|
| 1 | 6 |
| 2 | 4, 6, 9 |
| 3 | 5, 6, 8, 11, 13, 14 |
| 4 | 2, 5, 6, 9, 11, 13 |
| 5 | 2, 3, 4, 8, 9, 13 |
| 6 | 1 2, 9 |
| 8 | 3, 5, 13, 14 |
| 9 | 2, 4, 5, 6, 11, 13 |
| 11 | 3, 4, 6, 9, 13 |
| 13 | 4, 5, 11 |
| 14 | 3, 8 |

Notice that Subject 1 is never a recipient, and Subject 2 never an Initiator, unless we fudge a little on appreciability for these cases-- as we have done for this example for the circled Recipients.

We now test this simple sequential circulation scheme on 11 items, to see what happens. We have used the first 11 titles listed after the initial 110 used to get the $S_{ij}$ values.

Table 2

| Item # | If Initiated by Subject # | Sent first to Subjects # | Sent next to Subjects # | Finally sent to Subjects # |
|--------|---------------------------|--------------------------|-------------------------|----------------------------|
| 115 | 1* | 6 | | |
|     | 4* | 2, 5, 6, 9*, 11, 13 | | |
|     | 9* | 2, 4*, 5, 6, 11, 13 | | |
| 116 | 4* | 2, 5, 6, 9*, 11, 13 | | |
|     | 9* | 2, 4*, 5, 6, 11, 13 | | |
| 117 | 6* | 1, 2, 9 | | |
| 120 | 2* | 4, 6, 9 | | |
| 123 | 2* | 4, 6, 9 | | |
| 130 | 13* | 4, 5, 11 | | |
| 136 | 11* | 3, 4, 6, 9, 13 | | |
| 137 | 4* | 2, 5, 6, 9, 11*, 13 | 3 | |
|     | 11* | 3, 4*, 6, 9, 13 | 2, 5 | |
| 142 | 6* | 1, 2, 9 | | |
| 144 | 5* | 2, 3, 4, 8, 9, 13 | | |
|     | 11* | 3, 4, 6, 9, 13 | | |
| 145 | 1* | 6* | 2*, 9* | 4*, 5, 11, 13 |
|     | 2* | 4*, 6*, 9* | 1*, 5, 11, 13 | |
|     | 4* | 2*, 5, 6*, 9*, 11, 13 | 1* | |
|     | 6* | 1*, 2* 9* | 4*, 5, 11, 13 | |
|     | 9* | 2*, 4*, 5, 6*, 11, 13 | 1* | |

Table 3

| Title # | Initiated by Subject # | Sent first to Subjects # | Sent next to Subjects # | Not sent to interested Subjects # |
|---|---|---|---|---|
| 1 | 9 | 4*, 5, 6, 11* | 3 | |
| 2 | 9 | 4, 5, 6, 11 | | |
| 3 | 9 | 4, 5*, 6*, 11 | 1*, 3 | |
| 4 | 9 | 4, 5*, 6*, 11 | 1*, 3 | |
| 5 | 9 | 4*, 5, 6*, 11 | 1 | |
| 6 | 5 | 3*, 4, 9 | 6, 11 | |
| 7 | 11 | 3*, 4, 6, 9 | 5* | |
| 10 | 6 | 1, 9 | | 5, 11 |
| 11 | 6 | 1*, 9 | | 5, 11 |
| 12 | 6 | 1, 9 | | |
| 13 | 6 | 1, 9 | | |
| 14 | 6 | 1*, 9* | 4, 5, 11 | |
| 15 | 6 | 1, 9* | 4, 5, 11 | |
| 16 | 6 | 1, 9 | | 11 |
| 17 | 6 | 1, 9 | | |
| 18 | 4 | 5, 6*, 9, 11 | 1 | |
| 19 | 4 | 5, 6*, 9*, 11 | 1 | |
| 20 | 4 | 5, 6*, 9, 11 | 1 | |
| 21 | 4 | 5, 6*, 9*, 11 | 1, 3 | |
| 22 | 4 | 5, 6*, 9, 11* | 1, 3 | |
| 23 | 4 | 5, 6, 9, 11* | 3 | 1 |
| 24 | 4 | 5, 6*, 9, 11 | 1 | |
| 25 | 4 | 5, 6, 9, 11 | | |
| 26 | 4 | 5, 6, 9, 11* | 3 | |
| 29 | 5 | 3*, 4, 9 | 6, 11 | |
| 30 | 9 | 4, 5*, 6, 11 | 3 | |
| 31 | 9 | 4, 5, 6*, 11* | 1*, 3 | |
| 32 | 9 | 4, 5, 6*, 11 | 1 | |
| 33 | 9 | 4, 5, 6*, 11 | 1* | |
| 34 | 9 | 4, 5, 6, 11 | | |
| 35 | 9 | 4, 5, 6, 11 | | |
| 36 | 9 | 4*, 5, 6, 11* | 3 | |
| 37 | 6 | 1, 9 | | 5 |
| 38 | 6 | 1*, 9* | 4, 5, 11 | |
| 39 | 6 | 1, 9* | 4, 5, 11 | |
| 41 | 4 | 5, 6, 9, 11* | 3 | |
| 42 | 4 | 5, 6, 9, 11* | 3 | |

* Indicates a Recipient who judged the item as relevant.

It is interesting to note from Table 2 that every interested Subject gets title 145 regardless of who initiates it, which is good, but this is not so for titles 115 and 144.

The same scheme was next tried on 37 of the titles initiated by 7 of these Subjects (#1, 3, 4, 5, 6, 9, 11) after the DICO distribution began.    Subjects 10, 12, 17 and 26 were excluded because they were initiated by these four excluded Subjects.    The results are displayed in Table 3.

The final result of this very limited trial is summarized in Table 4.

Table 4

| Subject # | # Titles Initiated | Total # of Titles Rec'd | # Titles Rec'd & Relevant | # Titles Rec'd & Irrelevant | # Relevant Titles Missed |
|-----------|-----|-----|-----|-----|-----|
| 1 | 0 | 23 | 7 | 16 | 1 |
| 3 | 0 | 15 | 3 | 12 | 0 |
| 4 | 11 | 18 | 3 | 15 | 0 |
| 5 | 2 | 28 | 4 | 24 | 3 |
| 6 | 11 | 26 | 12 | 14 | 0 |
| 9 | 12 | 25 | 6 | 19 | 0 |
| 11 | 1 | 29 | 9 | 20 | 3 |
| Totals | 37 | 164 | 44 | 120 | 7 |

This seems to be quite satisfactory, on the whole, since over one-fourth of the titles received were judged to be relevant; rather few titles were missed, actually only 7 out of 51.    This completes the description of the trial of this simple sequential circulation system.

SASIDS Design

We go first to a description of a stochastic adaptive sequential information dissemination system (SASIDS) and then make some comparisons between DICO and SASIDS.

The description starts with the assumption that a Participant i is able to assign a value $0 \le r_{ij} \le 1$ that measures the degree of relevancy of title j in his work, at least at the time of reading. His rating is expected to be the same whether or not he has already seen Abstract j or the item to which it relates. For simplicity of exposition, it is assumed that titles are numbered serially as they are introduced into the SASIDS system.

Actually, SASIDS uses titles plus Abstracts instead of titles only, as does DICO. SASIDS creates a set of probabilities $P_{ik}{}^{j}$ to use in determining whether or not Abstract j judged by Participant i to have relevancy index value $r_{ij}$ will or will not be sent to Participant k, because of this feedback to SASIDS of $r_{ij}$ by Participant i. Specifically, SASIDS sends Abstract j to Participant k with probability $(r_{ij}P_{ik}{}^{j})$. Similarly, and in a stochastic sequential manner, if SASIDS does in fact transmit Abstract j to Participant k then SASIDS next sends Abstract j to Participant $\ell$ with probability $(r_{kj}P_{k\ell}{}^{j})$. Of course, Abstract j would not be transmitted physically to Participant $\ell$ if SASIDS had previously done so during the stochastic sequential circulation process, but the decisions regarding further circulation beyond Participant $\ell$ would be made as though this $r_{\ell j}$ had not been used previously.

The adaptive behavior of SASIDS results from the way in which the values of $P_{ik}{}^{j}$ are modified after a transmission of Abstract j to Participant k occurs and the value of $r_{kj}$ is available. One among many possible rules for such a modification is given by the following equation:

1)      $\overline{P}_{ik}{}^{j} = u_{k}P_{ik}{}^{j} + (1-u_{k})r_{kj},$

where $0 \le u_{k} \le 1$ is a parameter yet to be selected and personal to Participant k. If we assume, again only for simplicity of exposition, that the entire sequential circulation of Abstract j is completed before Title (j+1) enters the system then $\overline{P}_{ik}{}^{j}$ is the new value taken by $P_{ik}{}^{j}$ immediately following the transmission of Abstract j to Participant k, because of the reaction of Participant i to it on this occasion. Of course, Abstract j could be transmitted again later to

Participant i and result in another modification of his transmission probabilities although Participant i actually receives Abstract j just once.

All that remains needed to specify the SASIDS system is a procedure for selecting the values $u_k$ and $P_{ik}^0$. For example, we

might arbitrarily set $u_k=0.8$ and $P_{ik}^0=1$. Or, we might equally well allow Participant k to select a value for $u_k$ and change it at any time he may wish to do so.

Let us turn now to consider the behavior of a system in which $u_k=0.8$ and $P_{ik}^0=1$.

## SASIDS Behavior

One way to display the dynamic characteristics of SASIDS is to consider the effect on a $P_{ik}^j$ of a feedback $r_{kj}$ for each of sever-

al typical values for $u_k$, $P_{ik}^j$, and $r_{kj}$, as is done in Table 5. We

note that $\overline{P} > P$ if and only if $P < r$, and more generally that the value of $\overline{P}$ always falls in the interval between P and r. At the extreme values for u, the value of $\overline{P}$ is either P or r and neither of these extremes seems useful. The case u=1 would not take any account of the feedback value for r, and so the system would not be adaptive. The case u=0 would ignore the past history of the system, and would simply lead to a distribution that terminated permanently as soon as a value r=0 occurred. In between these extremes, we see that high values of u cause slow but rather steady adaptation whereas a low value like u=0.1 causes rapid and quite unsteady adaptation.

Perhaps u=0.8 or u=0.9 would be good values for this design parameter for a Participant whose history in the system is a long one and whose interests have not changed greatly, whereas a Participant whose interests have changed might prefer to set u rather small until such time as his more recent history causes the system

Table 5

| $u_k$ | $P_{ik}{}^j$ | $r_{kj}$ 0 | 0.1 | 0.25 | 0.5 | 0.75 | 0.9 | 1.0 |
|---|---|---|---|---|---|---|---|---|
| | | | | Values of $\overline{P}_{ik}{}^j$ | | | | |
| 0.9 | .1 | 0.09 | 0.1 | 0.115 | 0.14 | 0.165 | 0.18 | 0.19 |
| | .2 | 0.18 | 0.19 | 0.205 | 0.23 | 0.255 | 0.27 | 0.28 |
| | .5 | 0.45 | 0.46 | 0.475 | 0.5 | 0.525 | 0.54 | 0.55 |
| | .8 | 0.72 | 0.73 | 0.745 | 0.77 | 0.795 | 0.81 | 0.82 |
| | 1.0 | 0.9 | 0.91 | 0.925 | 0.95 | 0.975 | 0.99 | 1.0 |
| 0.8 | .1 | 0.08 | 0.1 | 0.13 | 0.18 | 0.23 | 0.26 | 0.28 |
| | .2 | 0.16 | 0.18 | 0.21 | 0.26 | 0.31 | 0.34 | 0.36 |
| | .5 | 0.4 | 0.42 | 0.45 | 0.5 | 0.55 | 0.58 | 0.6 |
| | .8 | 0.64 | 0.66 | 0.69 | 0.74 | 0.79 | 0.82 | 0.84 |
| | 1.0 | 0.8 | 0.82 | 0.85 | 0.9 | 0.95 | 0.98 | 1.0 |
| 0.5 | .1 | 0.05 | 0.1 | 0.175 | 0.3 | 0.425 | 0.5 | 0.55 |
| | .2 | 0.1 | 0.15 | 0.225 | 0.35 | 0.475 | 0.55 | 0.6 |
| | .5 | 0.25 | 0.3 | 0.375 | 0.5 | 0.625 | 0.7 | 0.75 |
| | .8 | 0.4 | 0.45 | 0.525 | 0.65 | 0.775 | 0.85 | 0.9 |
| | 1.0 | 0.5 | 0.55 | 0.625 | 0.75 | 0.875 | 0.95 | 1.0 |
| 0.3 | .1 | .03 | 0.1 | 0.205 | 0.38 | 0.555 | 0.66 | 0.73 |
| | .2 | .06 | 0.13 | 0.235 | 0.41 | 0.585 | 0.69 | 0.76 |
| | .5 | .15 | 0.22 | 0.325 | 0.5 | 0.675 | 0.78 | 0.85 |
| | .8 | .24 | 0.31 | 0.415 | 0.59 | 0.765 | 0.87 | 0.94 |
| | 1.0 | .3 | 0.37 | 0.475 | 0.65 | 0.825 | 0.93 | 1.0 |
| 0.1 | .1 | .01 | 0.1 | 0.235 | 0.46 | 0.685 | 0.82 | 0.91 |
| | .2 | .02 | 0.11 | 0.245 | 0.47 | 0.695 | 0.83 | 0.92 |
| | .5 | .05 | 0.14 | 0.275 | 0.5 | 0.725 | 0.86 | 0.95 |
| | .8 | .08 | 0.17 | 0.305 | 0.53 | 0.755 | 0.89 | 0.98 |
| | 1.0 | .1 | 0.19 | 0.325 | 0.55 | 0.775 | 0.91 | 1.0 |

to settle down with respect to his altered relevancy criterion.

It should be remembered that many other adaptation procedures could be used equally well. Also, many variations on SASIDS
as proposed are worth attention. For example, the probabilities
could be calculated as proposed but distribution of Abstract j might
always be made to Participant k when $P_{ik} \geq 0.7$, on the reasonable
assumption that a linkage from i to k with this high probability value indicates that Participant k is sufficiently likely to be interested
in Title j to justify sending it to him.

We next discuss a few important variations that should be considered seriously for any particular application of SASIDS.

SASIDS Variations

The SASIDS system was described as though the Participants
were in fact all different persons, and this would indeed sometimes
be a desirable design. Usually, however, it will be preferable to
encourage each person to constitute himself as several independent
Participants. In this way, each person can reflect his own widely
different fields of interest through the several Participants constituting him. For example, a person might use one of his Participants
for his field of scientific research, another for his hobbies, another
for his administrative position, and so on.

Instead of Titles, it may be preferable to exchange Abstracts.
The Abstract need not be treated as an impersonalized brief summary of a document of some kind, and also as though only one Abstract would normally be introduced into SASIDS to represent one
document. Actually, the Abstracts need not refer to documents but
could be any kind of object whatsoever. In particular, it would be
highly desirable for each Participant to introduce into SASIDS whatever items he believes to be useful for the purpose. However, for
our present purposes, we shall assume that the items introduced into SASIDS are all graphical in form and easily reproducible for distribution purposes.

It will nevertheless usually be highly desirable for any Participant to introduce another Abstract for any document that he feels

is quite relevant to his interests in some way not covered adequately
by previous SASIDS Abstracts he has seen, including any Abstracts
for the same document that he may himself have introduced previously.  Indeed, it should be common for one person to introduce Abstracts from two or more of the Participants constituting himself;
for example, an article on management science might contain:  1)
a mathematical result of relevance to the person in his Participant
role as a research mathematician, 2)  significant managerial guidance for him as another Participant in this administrative role, and
3)  useful formulas or numerical tabulations relevant for him as another Participant whose hobby is chess.  This multi-Participant feature makes possible many variations other than those already discussed, but we shall now leave this topic to the reader.

    The multi-Participant feature might be developed in quite a dif-
ferent manner, where SASIDS creates and destroys the constituent
Participants for each person without his direct knowledge or con-
cern--in contrast to the preceding method which requires that each
person select, define, maintain, and revise his own several Participant roles.  If a broad classification of subject-matter fields is constructed such that each Abstract can be identified with one or more
of these Classes then a Participant establishes his current interest
in a Class by returning high relevancy index values for Abstracts he
rates that are identified with this Class, and the Participant would
then be reconstituted as several Pseudo-participants representing
each of these Classes.  The Pseudo-participants would be treated by
SASIDS as though they were actual Participants but the person involved would never be concerned with anything but stating relevancy
for each Abstract and without concern for, nor knowledge of, the
nature of the Pseudo-participants that constitute him.  Pseudo-participants would also be taken out of the system when the average
relevancies attributed to them became too small.  Obviously, the
record-keeping necessary for a Pseudo-participant system of this
kind would be too great to manage unless the number of persons included under SASIDS were limited, and an adequate computing facil-

ity for such purposes might well be too costly to justify the effort.

As described, SASIDS depends upon the persons using it to introduce Abstracts that will be of value to others as Participants. It would also be possible to introduce Abstracts in many other ways. For example, each person might also provide SASIDS with a set of descriptors delimiting his current reading interests and all items in a set of standard abstracting periodicals could be screened by SASIDS to select those matching a descriptor-set for any Participant, with the Abstract going first to those Participants with such a match. This method has the great advantage that the Participants need not take the initiative in selecting and introducing the Abstracts, and the equally great disadvantage that the flow of Abstracts to Participants might quickly overload them.

Another type of variation is one that enables Participants to forestall transmission of Abstracts that they know in advance they do not wish to see. For example, SASIDS could easily afford Participants the option to list periodicals, or subjects, or authors, or publication dates, that identify work they prefer not to receive. Or, Participants could make positive requests of SASIDS for types of Abstracts similarly defined. Again, refined options of this kind would greatly increase the complexity of SASIDS, and its cost, but the added benefits to Participants will sometimes justify such refinements.

Selection from among variations of SASIDS, like those discussed in this paper, is probably best made on the basis of experimentation with small groups of Participants. Experimental trials of this type are themselves difficult to conduct effectively, for it is always very hard to devise a satisfactory measure of the relative performance of alternative information dissemination systems. It is even harder to compare benefits with costs for any such system, but these matters will not be discussed further here.

<h2 style="text-align:center">Notes</h2>

1.    University of Michigan and Consultant, IBM Corporation.

2.   Kochen, M. and Wong, E., ''Concerning the Possibility of a Cooperative Information Exchange,'' IBM Journal of Research and Development, 6, No. 2, April 1962, pp. 270-271.

## V.  Conclusion

What unifies the material in the preceding four chapters is
the convergency, by a number of theoreticians from different disciplines, backgrounds, viewpoints, on a common central problem:
how can relevant information in our increasingly complex environment be represented, stored/recalled, and processed, communicated
efficiently so that increasingly effective actions can result.  Another
unifying factor for some of the work is the exercise of specifying and studying the AMNIP system.  This is based on: the representation of information in a formal language built of names
and predicates which is extendable so as to become increasingly
English-like; on storage and recall of items represented as lists of
such sentences and modelled as a graph with names as nodes and
edges as predicates (or relations) between names; on processing as
a search over the trails of this net and grouping of "similar" nodes
into larger aggregates.

Thinking in terms of the AMNIP system has made it possible
to bring a fresh approach to the construction and use of thesauri--
one that may be more practical and effective than contemporary and
older methods--to the use of citation nets, and to bibliographic
classification methods.

On the above-mentioned central intellectual problem, some novel ideas of how internalized information can be represented (stored
and processed) as patterns built up from continually entering raw
"sense" data have been developed and are being tested by simulation.  On the whole, however, the problem presents as great a challenge as ever, perhaps even more so now that it is somewhat better
understood.

Some of the results in the separate papers are interesting in their own right.   Work by W. D. Frazer on the problem of assigning hardware images to words describing (as multiple, coordinate descriptors) a file of records, then searching first the hardware images and next all words under one image led to the following finding.   If the assignment of descriptor words to records is random (as it is for automobile license numbers, etc.), then an increase in hardware by a factor of $1/n$ results in an increase in search efficiency by a factor of n.

On the question of reorganizing memory so as to continually improve search efficiency as memory grows, C. Abraham was able to apply graph theory to group (cluster) items which share enough attributes with each other.   Such items might be words linked to other words as in a thesaurus.   This has led to programmed algorithms for finding "clusters."   Several important practical by-products, such as the application of the cluster algorithms to minimizing the interconnections in electrical assemblies, resulted.   Further practically useful fallout from these investigations includes the operational IPL-V system on the IBM 7040 and a preliminary analysis for the evaluation of certain document retrieval systems.

The novel dissemination systems DICO and SASIDS have been implemented as realistic experiments.   The latter is now--at the time of publication a widespread experimental system which shows promise of being operational and useful.

Much of the material, however, is speculative and discursive. While a definitive, mathematical treatment is eventually needed, it is better to admit that we are still in a stage of searching for the right problems, for the needed foundations at a verbal level than to pretend that all this is already done by premature mathematization.

Thus, we have portrayed what may well be the current state of theory in information science: intellectual ferment in the discovery of new concepts; search for definition and classification of basic issues; explorations into the logical structure of information processes like representation, storage/recall; an emerging capability to formulate and solve problems at a reasonably high level; maturation of judgment about what is fundamental and realistic in the analysis and design of information systems.

# Index

ABRAHAM, C. T., 128, 129, 131, 149, 171, 172, 174, 225, 229, 230, 252, 264, 270, 290

Abstraction, process of, 76
See also: Generalization

Absurdity (Sociology), 93

Acceptance-rate, 49, 54, 58, 120, 140, 196, 197, 198, 200, 271, 274, 277, 280, 285
See also: False drops; trash; relevance rate; efficiency

Access-currency-detail trade-off, 158

Access time (computers), 49
See also: Systems analysis

Accounting (Sociology), 87

Action-line, admissible (Sociology), 86

Action routines for computer learning, 111, 113
See also: Computer programming; learning

Action theory (Sociol.), 68, 69, 79

Action theory, in terms of state transition graphs, 70

Actor, definition of, 79

Adaptive reorganization of thesaurus graph, 266, 270
See also: Thesaurus

Adaptive procedures, 21, 266, 270, 271, 276, 282, 285
See also: SASIDS, Learning, AMNIPS

Addresses, of adjacent memory cells, 189

Addressing, open, 208
See also: buckets; blocks; Computation time

Addressing of memory, 154, 189, 195, 207, 208, 212

ADI - American Documentation Institute, 41, 130, 159, 171

Adjacency matrix, 181

Aesthetic (Sociol.), 87, 91

AFCRL - Air Force Cambridge Research Laboratories, 14, 25, 41, 83, 105, 128, 156, 171, 205, 215, 222, 251, 264, 270, 275

AFOSR - Air Force Office of Scientific Research, 156, 172, 230
See also: WOOSTER, H.; SWANSON, R.

293

295

ology), 75

CORNELL UNIVERSITY, 67, 83

Costs, of IR system, 49, 53, 55, 56, 58, 59

See also: Systems analysis

Criterion function, in addressing, 214

Critical constant, of crystal in percolation process, 176

Cross-correlation, between fields in addressing, 203

Cryogenics, for associative memories, 220

Crystal, definition in percolation process, 175

Current awareness, 27

See also: Dissemination; DICO; SASIDS, Alerting, Intelligence

CURRENT CONTENTS, 271, 272

See also: Current Awareness

Cycle connectivity, in graphs, 133, 231

D

Data fields; in addressing, 187

Data images, functional, 204

See also: Addressing

Data-processing, 16, 225

Data retrieval, 187

See Information Retrieval

DAVIES, P., 221, 222

DAVIS, W., 159

DAWE, A. R., 275

Decision theory, as related to orientation theory, 70, 79, 81

Decomposition of graphs, 231

Decomposition of undirected bipartite graph, algorithm, 248

Demand-curve, for IR service, 53, 58

Descriptors, 48, 187

DICKENSON, A. H., 171, 172

DICO - Discovery (or dissemination) of Information through Cooperative Organization, 227, 271, 273, 276, 277, 290

DICO, hit-rate, 280

DICO, experimental results on acceptance-rate, 280

Discourse, unlimited domains of, 94

Discourse, systems of, 67, 74

Discourse analysis, 62

Dissemination, 126, 271, 272, 274, 276

See also: DICO; SASIDS; Alerting; Intelligence; Current Awareness; SDI

Display, terminals, 165, 266

See also: Consoles; input-output

Distance, between nodes in thesaurus graph, 267, 135

See also: Graphs; Thesaurus

Document, as a cultural artifact, 227

Document-document coupling, 42

Document-retrieval, 47

HARRIS, T. E., 186

HAYES, R. M., 205

Heuristics, 11
See also: Learning; Intelligence; Theorem-proving; Inference; Hypothesis-formation; Cognition

Hierarchies, 131, 134, 135, 230, 234
See also: Classification

Hit-rate, 49, 54, 58, 120, 140, 164, 198, 271, 280
See also: Effectiveness, Recall rate; miss

HOLLAND, J., 221

HOLM, B. E., 129

HOLZNER, B., 84

Homographs, 96, 118

Homomorphisms, on graphs, 245

HORTY, J. F., 129

HUNT, E., 110

Hyphenation by computer, 34

Hypothesis-formation, 11, 33, 102
See also: Concept formation

I

IBM 7040, 113, 290

IBM 7090-4, 13, 113, 116, 201, 213

IBM patent department, Poughkeepsie, 165

IBM RESEARCH CENTER, 14, 83, 105, 273, 287

Idioms, 102, 108

Impotence (Sociology), 93

Inbreeding, in citation nets, 163

Incidence matrices, 255

Index terms, clusters, 53

Index terms, hierarchy, 234

Indexing, 59, 65, 127
See also: Coordinate; Cataloging; Permuted title; thesaurus; Search

Indexing and citation, 161
See Citation indexing

Indexing inconsistency, 127

Indexing Language, design of, 118, 120, 121

Indicators, in alerting, 32

Individuation (Sociology), 91

Inference, by computer, 31, 32, 98, 101, 225
See also: Heuristics; Theorem-Proving; Cognition; Learning; Hypothesis-Formation; Clustering

Information
See Knowledge; Coding

Information "crisis," 156

Information flow patterns, 43

Information retrieval, 12, 174, 187
See also: Document Retrieval

Information Science, as emergent discipline, 14, 19.

Information systems, 41, 42, 61

Infringement, in patent searches, 169

Input-Output, improved facilities, 97, 113

See also: Terminals; Display

Instruction, computer-based

See Teaching

Integration (Socio-Psychol.), 93

Intelligence, Artificial, 11, 12, 20, 30, 31, 32, 43, 82, 103, 105, 109

Intelligence, military, political, etc., 33

See also: Alerting; Dissemination.

Integer linear programming, relation to addressing, 211

Invisible colleges, 16, 44

See also: Cliques; Clusters

IPL-V, 21, 94, 113, 114, 115, 116, 217, 290

See also: Computer programming

IPL V, multiprogramming feature of, 114

Irreducibility, in graphs, 245

Irrelevancy, 271

Irritation (Sociology), 93

Isolation (Sociology), 93

J

JAMES, W., 79

Judgment (Sociology), 85

JUNG, R., 67, 83

K

KEHL, W. B., 129

KELLER, H., 111

KELLY, ., 68

KESSLER, M. M., 172

Keywords, 153

Kev-address transformations, 210, 214

Key patents, 164

KING, G. W., 160

KISEDA, J. R., 223

KLAUSNER, S., 83

Knowledge, 46, 61, 62, 80,

See also: Comprehension; Learning; Cognition; Concept-Formation

KOCHEN, M., 14, 25, 41, 47, 60, 61, 63, 94, 102, 104, 123, 129, 151, 156, 160, 163, 171, 172, 227, 230, 264, 265, 270, 271, 275, 276, 288

KÖNIG, D., 236

L

Language learning, 64, 102, 107

LAWLER, E. L., 151, 259, 264

Leaf decomposition, computer program use instruction, 134, 232, 235, 251

Leaf decomposition, use of, 136 256

Learning, 65, 76, 94, 101, 109, 111, 112, 113, 265, 275

Leaves, in graph, example, 137, 231, 233, 234, 243

301

LEE, C. Y. , 221, 223
LEE, E. S. , 223
LEEDS, A. , 83
LEWIN, K. , 68
LEWIN, M. H. , 220, 224
Lexicon, 117
   See also: Vocabulary, The-
saurus
LIBBY, R. L. , 160
Library automation, 12, 160
Library of Congress, automa-
tion, 160
Limited terms, in queries, 124
   See also: Meaning contractors
LIN, A. D. , 216
Linear programming, related
to addressing
   See Integer
Linguistics, 11, 48, 49, 60,
65, 66, 96, 106, 116, 117,
118, 119, 123, 124, 125,
131, 132, 136, 138, 141,
229, 266
   See also: Generative Gram-
mar; Vocabulary
Linguistic performance, 108,
116
LIPETZ, BEN-AMI, 172
Lists, 217
Lists, comparison of, 51
List-processing, 64
List, structures, 114, 217
Literature, assimilation of, 156
LITTLE, A.D. , 172
Lobe decomposition, algorithm,

133, 136, 235
Lobes, in electrical assemblies
and graphs, 232, 233, 234,
258
Logic Gates, 252
LOS ANGELES TIMES, 34
LUHN, H. P. , 275

M

MAC, 40
   See also: Time-sharing
MACKAY, D. C. , 104
Magnetics, in associative memo-
ries, 220
Management, 26
MANDELSTAM, S. , 84
Man-Machine symbiosis, 21, 117,
123, 141, 163, 174, 180, 265,
266
MARON, M. E. , 104
Masking, 221
Mastery (Sociology), 93
Matching, operation in automata,
97
Matrices, Boolean multiplication,
236 245
   See also: Boolean
MC CLUSKEY, E. J. , Jr. , 215
Meaning (Sociology), 93
Meaning extendors, 88
Meaning restrictors, 88
MEMEX, 12, 19, 227
   See also: BUSH, V.
Memory, 49, 97, 151, 207

See also: Storage, Addressing

Memory, associative

See Associative

Memory, Dynamic reorganiza-

tion, 227

See also: Storage; Graphs

Memory organization

See File organization

Memory, training of, 227, 266

See also: Adaptive Procedures

MENDELSOHN, N. S., 248, 264

Metaphor, 96

Methodological (Sociology), 87

MICHIGAN, UNIVERSITY OF,

94, 287

Microform, 37,

Microprogramming, 154

MILLER, G. A., 64

Minimum cover problem, 212

Miss, 120, 140

See Hit-rate

MIT - Massachusetts Institute

of Technology, 40

Models, internal, 64

See also: Cognition

Modules, 252

MONTREAL, UNIVERSITY OF,

35

MOOERS, C. N., 129

Moral (Sociology), 87, 91

Morphological relations between

words, 125, 266

Motivation theory, as related to

orientation theory, 70, 79,

81

Moving, as computer operation,

111

MULLER, D. E., 215

Multi-lists, 217, 218

See also: Associative Memory

Multiprogramming, 114

See also: Time-sharing

MUROGA, S., 193, 205, 206,

216, 252, 264

See also: BEATTY, J.

Mystery (Sociology), 93

N

Names, 230-289

Naming, as computer operation,

98, 111

NATIONAL BUREAU OF STAND-

ARDS, 25

NATIONAL SCIENCE FOUNDA-

TION, 126, 130

NATIONAL UNION CATALOG, 37

NEEDHAM, MRS. K.

See SPARCK, JANIS, K.

Negation operation, in matching,

221

Networks

See Graphs

New-look cognitive theories, 68

NEW YORK TIMES INDEX, 34

NEWELL, A., 105

NEWHOUSE, V., 224

Newspaper, personalized, 272,

275

O

Object generators, 88
Objectifying transformation, 75
OLDS, J., 84
On-demand, 37, 47
  See also: Time-sharing;
  Man-Machine symbiosis
On-line,
  See Time-sharing; Man-
  Machine symbiosis
Operational analysis, 47
  See also: Systems analysis
Operators, types of, 78
Ordering, partial, 231
ORE, O., 150, 264
Organism, types of, 83
Orientation, definition, 71
Orientation, disorders of, 82
Orientation, systems of, 67
Orientation, as an operator,
  72, 185
ORNE, M., 68

P

Pain (Sociology), 93
PALERMO, F. P., 216
Pandemic operators (Sociology),
  76
Parity-check addressing, 212
  See also: Addressing
Parallel processors, 217, 218,
  221, 222
PARSONS, T., 68, 84

Partial ordering, 217
Patents, as data base for cita-
  tion index 162
Path, in percolation process,
  175, 181
Path-tracing, 164, 217, 266
  See Trail searching
Pattern-recognition, 62, 98, 99,
  102, 110
  See also: Classification;
  Learning
PAULL, M. C., 221, 223
PENNSYLVANIA, UNIVERSITY
  OF, 25
Perception, simulation of, 112
Perception, as epistemic orienta-
  tion (Sociol.), 72
  See also: Cognition
Percolation processes, as models
  for memory, 153, 174
PERLIS, A., 218
Permeability, in percolation
  process, 177
Permuted-title indexes, 45
  See also: Indexing
Personnel file, 201
  See also: Addressing, Random-
  access files
PETERSON, W. W., 216
PETRICK, S., 215
Photostore, 13
  See also: Table lookup; Random-
  dom-access files
PIAGET, J., 68, 109
Planning, 27, 31, 95, 225

Pleasure (Sociology), 93

Polysemes, 118

Pragmatic (Sociology), 87

Predicates (Logic), 22, 58, 230, 289

Prescription (Sociology), 93

PRIBRAM, K., 64

Price per query, 49, 53, 56, 58

See also: Costs

Prime numbers in incidence matrix calculation, 239

Problem-solving, 15, 108, 151, 225

See also: Theorem-proving, Intelligence, Inference

Processing time per query, 52

See also: Systems analysis

Processor organization, diffuse, 204

See also: Central Proc. Unit

Programming

See Integer, Linear, Computer

Progressive covering, principle of (Sociol.), 258

Proscription (Sociology), 93

Psychopathology, 82

PRYWES, N. S., 224

Pushdown lists, 218

Q

Queries, 48, 50, 126, 158, 163, 199

See also: Boolean

Query, dual form of, 199

Query reformulation, 124, 266

See also: Man-machine symbiosis

Query terms, frequency of recurrence, 126

R

RADC - Rome Air Development Center (438L), 128, 171

RAJCHMAN, J., 220, 224

RAMAC, random access memory, 208

RAND, 63, 104

Random-access files, 36, 49

See also: Addressing; RAMAC

Randomization, in addressing, 206, 213

Randomness, in data base, 203

RAPOPORT, A., 84

RAVER, N., 216

RAY-CHAUDURI, D. K., 215, 216

Reachability matrix (Graph Th.), 134, 236, 256

Recall, 15, 151

Recall-rate

See Hit-rate

Reductionism (Philosophy), 110

Redundancies in literature, detection, 152

Redundancy, semantic, 119

See also: Homographs; Synonyms

306

SELYE, H., 35
Semantic diversity, 119
Semantic ideolect, 119
Semiconductors, in associative
 memory, 220
Separating edge, in graphs, 233
Separating vertex, in graphs,
 233
Serial access, 195, 203
Serial read rate, 49
Set abstractors (Sociology), 88
SHAW, R., 28
Shifting, computer operation, 97
SHILS, E., 84
Similarity measures, 229
SIMON, H. 105

Simulation, by computer, 201
 213
SNOBOL, a computer program-
 ming language, 94
Sociometry of knowledge, 161
Solidarity (Sociology), 93
SOMMERHOFF, G., 84
SPACE SCIENCE LABORATORY,
 U. of Calif., Berkely, 227,
 274
SPARCK-JONES, K., 130, 149,
 230, 264
 See also:  NEEDHAM, J.,
 MRS.
Splitting set, characterization
 (addressing), 213, 214, 215
Star-graph, in thesaurus graph,
 268

See also: Thesaurus; graph
Storage capacity, 49, 54, 58,
 269
 See also: Memory
Storage, balance against search,
 265, 270
 See also: Systems analysis
Storage allocation, dynamic, 219
Storage of image in fast-access
 memory, 204
Storing, as an operation, 98
String processing, 221
 See also: computer program-
 ming
Strings of bits (Addressing), 187
STROM, R., 105
Strong cycle connectivity (Graph
 Th.), 235
Subjectifying transformation (Soci-
 ology), 75, 88
Summary, concept of, 157
SWANSON, D. R. 130, 160
SWANSON, R.,
 See AFOSR
Switching circuits, relation to
 addressing, 209
Symmetry of relations, in graphs,
 231
Synonyms, 48, 49, 125, 131,
 136, 138, 141, 229
SYSKI, R., 50, 60
Systems analysis, 47, 230, 265,
 270
 See also: Operational analysis
SDC - Systems Development Cor-

poration, 37